Bioweapons

Bioweapons

The History, The Present and The Future

Stig S Frøland

Pen & Sword
MILITARY

First published in Great Britain in 2026 by
Pen & Sword Military
An imprint of Pen & Sword Books Limited
Yorkshire – Philadelphia

ISBN 978 1 03618 269 4

A CIP catalogue record for this book is available from the British Library.

Typeset by Mac Style
Printed in the UK by CPI Group (UK) Ltd, Croydon, CR0 4YY.

MIX
Paper | Supporting responsible forestry
FSC www.fsc.org FSC® C013604

The Publisher's authorised representative in the EU for product safety is Authorised Rep Compliance Ltd., Ground Floor, 71 Lower Baggot Street, Dublin D02 P593, Ireland.
www.arccompliance.com

For a complete list of Pen & Sword titles please contact:

PEN & SWORD BOOKS LIMITED
47 Church Street, Barnsley, South Yorkshire, S70 2AS, England
E-mail: enquiries@pen-and-sword.co.uk
Website: www.pen-and-sword.co.uk
or
PEN AND SWORD BOOKS
1950 Lawrence Road, Havertown, PA 19083, USA
E-mail: uspen-and-sword@casematepublishers.com
Website: www.penandswordbooks.com

Contents

Foreword

For thousands of years, Homo sapiens has utilised his impressive creativity to develop increasingly effective and deadly weapons. Today, we differentiate between conventional weapons and so-called ABC weapons – a collective term for nuclear, biological, and chemical weapons. These are also referred to as weapons of mass destruction because they can kill a large number of people. Biological weapons consist of living organisms, particularly microorganisms, but also insects and certain toxins produced by microbes or plants. The potential for mass destruction is even greater with some types of biological weapons than with nuclear and chemical weapons.

Biological weapons can be used in warfare, acts of terror, and various criminal contexts. A range of factors, including advancements in science and technology, make biological weapons a significant and growing threat today. It is therefore surprising that there is a striking silence about this in the mass media and the public at large. My intention with this book is to provide ordinary readers without specialised knowledge insight into this important but under-communicated topic.

The book offers an overview of the key aspects of biological weapons – the various types of such weapons, their characteristics and potential for harm, as well as the ways they can spread. Separate chapters describe the long and dramatic history of biological weapons from pre-modern times to the present day. Furthermore, it discusses how we can – and must – address the threat posed by biological weapons, which, in the worst-case scenario, could lead to a pandemic and ecological Armageddon.

I would like to thank my wife, Ellen, and my son, Carl, for their constructive reading of the manuscript. Carl has also contributed with research and useful input.

Introduction

Humanity and Weapons

When the first members of Homo sapiens – a seemingly unimpressive animal species – wandered out onto the savannas of Africa 200–300,000 years ago, they faced many dangers from their environment, including threats from other, far more fearsome species with much more intimidating muscles, teeth, and claws. Fearsome, now-extinct species such as the sabre-tooth tiger, cave bear, and giant snakes threatened our ancestors. The fact that our modern branch of humanity not only survived this struggle but eventually became the dominant species on our planet is, in fact, a like a fairy tale.

The key to understanding this success story is, of course, Homo sapiens' unique brain capacity, which has allowed us to master our surroundings. Central to this is humanity's remarkable ability to develop useful tools. Even the earliest representatives of Homo sapiens used tools made from stone. But even the ancestors of Homo sapiens – the species Homo habilis – 'the handy human' – which lived in south-east Africa a couple of million years ago, utilised such tools, albeit in primitive forms.

Over time, Homo sapiens developed increasingly advanced tools that have been crucial to human development and the growing control over nature. These were made from stone, bone, and wood, and for some thousand years, also from metal.

The dark side of humanity's technological agility is that it has also been used for destructive purposes – the development of weapons. Long before recorded history, our ancestors discovered that tools could also be used to harm both animals and fellow humans. Weapons became important in hunting but were also used in violent clashes with other people, such as battles with foreign tribes.

Over time, humans settled down, and larger societies, cities, and states emerged. More extensive violent conflicts and wars became common, which has characterised human history up to the present day. Homo sapiens has used its significant capacity for innovation to develop increasingly effective and deadly weapons.

In the terrifying arsenal of weapons that exists today, one type stands out clearly – biological weapons or bioweapons.

What are bioweapons? The common denominator for this extensive and highly varied type of weapon is that they are based on the use of living organisms, primarily microbes, but also toxins – poisonous substances – produced by microbes or plants. Additionally, insects can also be used as bioweapons.

Bioweapons can be used to attack humans, animals, and plants with the intention of killing or incapacitating relevant targets. Such weapons can be used in war, during terrorist attacks, and in 'ordinary' crime.

These weapons are strikingly little discussed in public. The reason for this may be that people do not believe that the use of such weapons is relevant today – because they are banned by international agreements and are subject to moral condemnation. Despite this, it is an uncomfortable fact that bioweapons not only exist and are being developed in today's world, but they also represent a serious threat that is likely to increase in the coming years.

Chapter 1

A Sinister Arsenal of Microbial Weapons – Characteristics and Mechanisms of Bioweapons

The most relevant bioweapons are based on living microorganisms. The microbial world that surrounds us is extraordinarily vast and is divided into several main groups that differ significantly from each other. The most important groups that are relevant as bioweapons are bacteria, viruses, and fungi. Within each of these groups, only a few microbes are capable of causing disease in humans, and only some of these are currently considered relevant bioweapons. Scientific and technological advances will likely lead to the expansion of this group in the coming years.

Among the many microbes that can cause severe infections in humans, animals, and plants, we have now – after extensive research over the last 75 years – identified 30–40 microbes that are particularly suited as bioweapons. It is not enough for a microbe to cause serious harm; several other conditions must be met for it to be used practically as a bioweapon.

What characteristics should a bioweapon have?

The microbe must likely cause infection with severe disease.[1] Some bioweapons have high lethality rates, which may be desirable in certain situations. Others may be effective if they can incapacitate the victims over a certain period without causing high lethality. The ability to spread from person to person can also be advantageous, as this would allow the effects to spread gradually. Some commonly used microbes lack this ability yet can still be effective.

Furthermore, it is a prerequisite that the microbe can be produced in large quantities and stored over time without losing its pathogenic capabilities. This is particularly important for use in warfare.

It is also crucial that the microbe is robust enough to withstand the stresses of the dispersal method that is relevant for use as a weapon. It must not become inactivated too quickly after dispersal due to environmental factors such as sunlight and temperature. With various technical measures in the application, it is possible to influence this.

In addition to these characteristics, the specific microbe's ability to induce fear and panic is emphasized. This is particularly relevant in terrorist attacks. From the terrorists' perspective, it will be advantageous if the damage inflicted leads to a significant strain on hospital functions and the healthcare system and can otherwise contribute to a desired destabilisation of society.

It is also advantageous if the individuals in the target group for a bioweapon do not have immunity to the relevant microbe, for example, due to vaccination. If effective medications against the microbe or toxin are also lacking, it would, of course, increase the damage inflicted. Ideally, it is desirable that the users of a bioweapon themselves have immunity, for example, due to vaccination. Thus far, we have had vaccines against only a handful of the microbes that have been considered as bioweapons. This is likely to change in the coming years due to advancements in vaccine technology.

How to use bioweapons

When using bioweapons, it is crucial to have an effective method for delivering the microbe or toxin that ensures the best possible spread in the target area. Depending on the specific bioweapon, a variety of technical systems are applicable.[2]

Aerosols – very small particles that remain airborne for extended periods and can spread over long distances – are considered the most effective method for disseminating bioweapons. Infection occurs when the aerosol is inhaled. Bioweapons can be spread in many ways as aerosols. This can occur very effectively with missiles, that is, self-propelled, guided projectiles, and with grenades and bombs that release the bioweapon-containing aerosol. Under favourable weather conditions, aircraft can spread aerosols and cause infections over several hundred kilometres. In smaller-scale attacks, such as terrorism, aerosols can be dispersed using simple spray mechanisms.

Contaminating drinking water sources with microbes or toxins can also be an effective method for spreading certain infectious diseases. One problem here is that the bioweapon often becomes overly diluted, resulting in diminished efficacy. Terrorists can also contaminate food with microbes, triggering epidemics in that manner.

Other dispersal methods are also possible. For example, microbes can be spread using insects, as we know from many naturally occurring infectious diseases, such as malaria and yellow fever. Among terrorists, the ingenuity applied in the dissemination of microbial weapons has been impressive. Individuals have, among other things, been attacked with injections of toxins and by spreading microbes using letters.

Finally, we may also encounter bioterrorists who spread deadly microbes through their own infection by contaminating others – a parallel to 'suicide bombers' who have detonated themselves, as has been seen many times in recent years.

The disease presentations caused by various bioweapons vary widely. In some cases, they can be frightening, but regardless, they are often difficult to recognise, and this can delay or hinder the implementation of targeted countermeasures.

In the following sections, we will examine some of the most relevant and feared bioweapons that have notably been subject to analysis concerning risks and preventive measures.[1, 2]

The Most Relevant Bioweapons

Several bacteria rank very high on the list of bioweapons that could be used in warfare and terrorism.[2, 3]

Anthrax Bacteria (*Bacillus anthracis*)

This bacterium has been regarded as the most feared of all bioweapons. Anthrax was the first infectious disease in which a bacterium was identified as the cause. As a naturally occurring infection, anthrax primarily affects herbivorous animals, both livestock and wild animals, but it can also cause severe infections in humans. In Western countries, anthrax is now very rare, but it remains common in developing countries.

Anthrax bacteria can, like certain other bacteria, enter a dormant phase known as spores. The spores are extremely resilient; they can withstand drought, heat, cold, and many disinfectants. They can persist for years in both soil and buildings. It is the spores that enter humans and animals and cause infection by transforming into active bacteria. In nature, infection often occurs through cuts and sores in the skin, but it can occasionally also happen through the mouth/intestine or via inhalation, resulting in pulmonary infection.

As a bioweapon, it is primarily anthrax spores that are relevant. When spores in aerosol form are inhaled, it leads to lung infection. Symptoms usually show up within just a few days. Untreated, lung infection is fatal in most cases, and even with antibiotics, the lethality rate is high. It is relatively straightforward to produce larger quantities of anthrax spores that can be aerosolised. On a larger scale, such dispersion can occur via aircraft. The use of drones for dispersion would also be effective. In smaller-scale situations, related to terrorism, relatively small amounts of spores in a letter, for example, could lead to infection and death.

Estimates have shown that an anthrax attack with spores, if spread by aeroplane over a city of 500,000 inhabitants, under favourable wind conditions, could result in 95,000 deaths and 125,000 illnesses.[4] Anthrax is not contagious between humans.

Plague Bacteria (*Yersinia pestis*)

Over the last 1,500 years, the plague bacteria have taken hundreds of millions of human lives. This has occurred through three pandemics – including the Black Death – along with numerous smaller epidemics. The lethality rate from infection is extremely high. It is thus not surprising that it occupies a place as one of the most feared bioweapons.

In nature, the plague bacteria are typically found in various rodent species, including rats. Humans can become infected through flea bites, as fleas can become carriers of the bacteria after biting host animals, such as rats.

Yersinia pestis can cause three different clinical manifestations. The classic form of the plague is bubonic plague, where the patient develops painful swellings (buboes) in the groin, armpits, and neck. Some patients develop a life-threatening lung infection – pneumonic plague. They can then infect others through droplet transmission, where tiny, microbe-containing droplets from respiratory secretions are inhaled by the victim. A third form, known as septicaemic plague, because the bacteria spread through the bloodstream, leads to rapid death without allowing the patient time to develop buboes or pneumonia.

The most effective use of plague bacteria as a bioweapon is to spread it as an aerosol, for example, by aircraft. A large number of individuals in the target area would then inhale the bacterial aerosol and develop pneumonic plague. Transmission could then occur further via droplet spread. Infected individuals could subsequently spread the plague across broader areas. The use of plague-infected fleas to spread bubonic plague is less relevant in the context of bioweapons, even though there are examples of this. It is much less effective than spreading the bacteria using aerosol methods.

Estimates have shown that 50 kilograms of plague bacteria spread as aerosol by aeroplane over a population of 5 million could cause pneumonic plague in 150,000 people. If appropriate antibiotic treatment is not initiated within a day, most will die.[4]

Tularaemia (Rabbit Fever)

The official name of the bacterium that causes what is called rabbit fever is *Francisella tularensis*. The medical term for the disease is tularaemia. The bacterium is found in nature, especially in various rodent species, including hares, and humans become infected in various ways from these animals. This can occur through direct contact with the animal, through inhalation of dust containing bacteria from animal faeces, through bites from ticks or mosquitoes, and through drinking contaminated water. In practice, the bacterium does not transmit from human to human.

It is believed that inhalation of just eight to ten bacteria is enough to cause illness, which can take many forms. Though the lethality rate is not particularly high, the high contagiousness makes the bacterium regarded as an interesting bioweapon, as a significant number of humans could potentially become ill and incapacitated.

This bacterium also spreads most effectively as an aerosol. Some 30,000 can die and 125,000 become ill in a population of 500,000 through inhalation of the aerosol spread by aircraft.[4]

Glanders and Melioidosis

The two bacteria *Burkholderia mallei* and *Burkholderia pseudomallei* are closely related and both are relevant as bioweapons.

Burkholderia pseudomallei is currently far more significant than its relative as a cause of severe disease in humans. The disease is called melioidosis. Melioidosis is relatively common in certain Asian countries and northern Australia but also occurs in Africa and in North and South America. In our part of the world, it is sometimes seen as an imported disease. The bacterium occurs in soil. People can be infected through skin contact, via the respiratory system, and through the mouth, including via contaminated drinking water. The infection affected American soldiers during the Vietnam War, gaining the name 'the Vietnamese time bomb' because it could remain completely asymptomatic for several years before suddenly manifesting with severe progression.

Burkholderia mallei is primarily an infection in hoofed animals, including horses. The infection is called glanders in English. It is now almost eradicated in most countries but is detected sporadically in Asia, Africa, and South America. It is highly infectious to humans through contact with infected animals but does not occur freely in nature.

The disease presentations caused by these two bacteria have many common characteristics. They are highly variable and can affect the lungs, skin, and many other organs. In the most severe forms of infection, the lethality rate is very high without antibiotic treatment.

Both *Burkholderia mallei* and *Burkholderia pseudomallei* are considered highly relevant bioweapons. The bacteria are relatively resistant to environmental influences, and they can be used as aerosols where infection in the target group occurs via inhalation of the bacteria. Further transmission from human to human is unlikely.

Germany used *Burkholderia mallei* as a bioweapon against enemy horses during World War I.

Typhus Fever

Typhus fever is often a deadly infectious disease caused by the bacterium *Rickettsia prowazekii*, which belongs to a specific subgroup of bacteria called *rickettsia.* This infectious disease has ravaged Europe for centuries but is currently very rare in our part of the world. Without antibiotic treatment, the infection has a lethality rate of up to 60%. Symptoms include fever, headache, muscle pain, a specific rash, and eventually confusion and unconsciousness. The bacterium is relevant as a bioweapon.

While typhus normally spreads through body lice that carry the bacteria, it is as an aerosol that it is most relevant as a bioweapon; producing aerosol with the bacteria is relatively uncomplicated.

Bacteria that Spread Through Drinking Water and Food

While the bacteria mentioned above primarily target their victims through the respiratory system in aerosols, some bacteria that spread by ingestion of contaminated water or food can also be utilised as bioweapons. This includes bacteria from the Salmonella group, such as *Salmonella typhi*, which causes typhoid fever – a severe infection. Bacteria from the Shigella group, which cause severe intestinal infections known as dysentery, are also relevant as bioweapons. The same goes for cholera bacteria – *Vibrio cholerae* – which has caused seven pandemics with high lethality rates over the last 500 years. Cholera continues to exist in many parts of the world, especially in countries with poorer sanitation and water hygiene than the standard in Western countries. It is also in these countries that cholera bacteria can be effective as military bioweapons. In our part of the world, the bacterium is likely most relevant for terrorists. There are several examples of this.

Viruses can also be used as bioweapons. Whether these very simple microbes should be called living organisms is a matter of definition. Viruses can only reproduce inside living cells. Among the enormous variety of viruses found in nature, only a few cause infection and disease in humans. However, some of these are responsible for some of the deadliest epidemics that have struck humanity throughout history. This includes smallpox virus.

The Smallpox Virus

The smallpox virus in its most dangerous form, called *Variola major*, took 300 million human lives during the twentieth century. It is not surprising that this virus ranks highly on the list of probable bioweapons. As a naturally occurring infection, smallpox has been eradicated in the world today. The virus, however, is preserved in frozen form in two research laboratories – one in the USA and one in Russia.

The virus spreads from human to human primarily through droplet transmission, where virus-containing respiratory secretions from patients are inhaled. The virus is highly resilient and can survive for extended periods in dried form outside the body in clothing and dust. Such dried virus particles can spread as aerosols. These characteristics make the virus in aerosol form highly suitable as a bioweapon. It is estimated that inhalation of just ten to 100 virus particles is sufficient to cause infection with the smallpox virus. The mortality rate from infection with *Variola major* is about 35%.

Currently, there are no well-tested medications against the smallpox virus, but one medication is promising (tecovirimat).

Viruses that Cause Haemorrhagic Fever

A number of viruses cause severe infections where bleeding from the skin and mucous membranes is part of the febrile, often dramatic clinical picture. Such viruses are categorised under the term haemorrhagic fever viruses. Several of them could be considered for use as bioweapons.

This includes, among other things, the so-called filoviruses, with the Ebola virus having caused extensive epidemics in Africa with a lethality rate of 50–90%. The same applies to its relative, the Marburg virus. Both of these filoviruses and other haemorrhagic fever viruses, such as various Arena viruses, are of interest as bioweapons. Again, it is aerosol dissemination that is relevant. These viruses are less stable than, for example, the smallpox virus. Addressing this problem requires special expertise.

Filoviruses can also, to some extent, transmit further from infected individuals through contact with infectious body fluids.

Currently, there are no effective medications against these viruses.

Foot-and-Mouth Disease Virus

The foot-and-mouth disease virus belongs to the so-called picorna family. The virus affects cloven-hoofed animals and pigs and is one of the most contagious viruses known. It can cause large epidemics with enormous economic losses. Hence, it can be used as a bioweapon to disrupt agricultural production.

Toxins from Living Organisms

Toxins – poisonous substances – produced by living organisms are also traditionally regarded as bioweapons, although some argue that they should be classified among chemical weapons. Several toxins are particularly relevant.

Botulinum Toxin – a Terrifying Poison

This toxin is produced by the bacterium *Clostridium botulinum*, which occurs in soil and water. It is a nerve toxin that leads to paralysis and is the deadliest of all naturally occurring toxins. Death is most often due to suffocation caused by paralysis of the respiratory muscles. Today, the toxin is known as part of the Botox preparation used in the cosmetics industry to remove wrinkles. Botulinum toxin has, moreover, a far grimmer history as one of the first agents that was attempted to be used as a bioweapon. This toxin remains high on the list of possible bioweapons.

Potential applications include either as an aerosol where the toxin is inhaled by the victims or by adding it to food or drinking water. It has been calculated that one gram of the toxin spread as an aerosol could kill a million people. Adding the toxin to drinking water reservoirs is likely less effective, but such use is nonetheless conceivable.

Ricin – an Effective Weapon of Murder

The toxin ricin can be relatively easily produced from the seeds of the castor bean plant *Ricinus communis*, which is readily available, including via the internet. Ricin is one of the deadliest toxins in nature and was therefore early on of interest as a bioweapon. It is deadliest when inhaled through the lungs. However, the lethality is also high via injection and after ingestion of food or drink. Death usually occurs within two to three days.

Ricin can be produced as an aerosol, but it requires such large quantities of the toxin when spreading from the air that it is considered unsuitable for attacks on larger populations. It is in small-scale attacks related to terrorism or murder attempts on individuals that ricin is relevant. There are many examples of this.

Toxins from Fungi – Mycotoxins

Living fungal organisms are not particularly relevant as bioweapons. However, several species of fungi produce so-called mycotoxins that are interesting in this context. In particular, the so-called T-2 mycotoxin is relevant. A few milligrams of this toxin can be lethal. It can be absorbed through the lungs, through the mouth, and, as the only known bioweapon, also through intact skin. Several states have produced mycotoxins as bioweapons, and they may have been used – but this is uncertain. In any case, these toxins belong on the list of potential bioweapons.

In addition to the mentioned examples of toxins, there is a long list of other toxic substances found in nature that might be expected to emerge as bioweapons.[3]

Insects – 'Six-Legged Soldiers'

In discussions of bioweapons, it is often forgotten that insects can also be used as weapons. This is relevant in two ways.[5]

Firstly, insects can be used for direct attacks on humans, animals, or plants, both in warfare and in terrorism. We already have examples of this from antiquity. Today, the most relevant use of insects is likely to destroy crops.

Secondly, insects can be used as carriers – 'vectors' – for microbes that can then attack humans and animals. In nature, this is a very important mechanism for a wide range of serious infectious diseases, with plague, malaria, and yellow fever being particularly well-known examples. As we have seen earlier, there have been attempts to spread the plague bacteria as bioweapons with infected fleas.

We now know that several states that have sought to develop bioweapons have conducted extensive trials with insect dispersal from aircraft. Biotechnological advances may also provide opportunities for genetic manipulation of insects that could increase their effectiveness as bioweapons.

The examples of biological weapons mentioned above represent only a small fraction of the possibilities based on well-known organisms and toxins. The microbial world is vast. Over recent decades, numerous new microbes have been discovered that, in many cases, can cause severe infections – so-called emerging infectious diseases. There is reason to believe that many of these newly discovered threats will eventually also be assessed as potential bioweapons.

Even more importantly, advancements in biotechnology will likely lead to entirely new bioweapons – partly genetic modifications of existing microbes and partly entirely newly produced microbes. Therefore, the list of potential bioweapons is likely to look very different in a few years compared to today.

Chapter 2

Snake Venom, Rats, and Rabies – Biological Warfare from Antiquity to 1900

It was not until the twentieth century that biological weapons were recognised as a real threat during military conflicts. However, the use of such weapons is not a new invention but can be traced back thousands of years. At first glance, this is surprising. Our knowledge of microbes as causes of infection and disease, and their ability to transmit from person to person, is quite recent. Modern understandings of microbes and infection emerged with what we call the bacteriological revolution at the end of the nineteenth century, with a number of pioneering researchers, most notably Robert Koch in Germany and Louis Pasteur in France. How could then our distant ancestors, who were unaware of the existence of microbes, utilise them for biological warfare?

Infection diseases in various forms, along with sickness and death, have always been part of human existence. From the moment the first humans stepped onto the African savannas, they had to grapple with the causes of these diseases. Religious explanations have often had a strong grip on people. Many believed that these diseases, which often manifested as epidemics, were sent by gods or demons as punishment for human behaviour.

But non-religious theories also gradually emerged. This includes the miasma theory, which was first proposed by Hippocrates in the fifth century BC and persisted among many doctors and scholars well into the nineteenth century. It was believed that the diseases we now know to be infectious were caused by disease-causing mists in the air – miasmas – that came from stagnant water, unhealthy swamp areas, and decaying carcasses of animals and humans. The Greek word 'miasma' actually means pollution.

The breakthrough for modern theories of infection and the belief in the role of bacteria did not happen overnight. In the mid-nineteenth century, there was a long-standing and partly fierce conflict between opponents and supporters, which only subsided towards the end of the century.

However, many infections are highly contagious. Therefore, it is surprising that this recognition did not emerge in the scholarly theories of disease. However, there is every reason to believe that the general populace, through their own experiences, was aware that certain diseases transmitted from person to person, even without understanding why. This is clearly supported by the Greek historian Thucydides, who described the terrible epidemic that ravaged Athens in the fifth century BC. He obviously assumed that the disease was transmitted from person to person, and his account shows that this was a common belief among the Athenians.

In humans, we find fundamental psychological mechanisms of disgust and revulsion towards things that we now know can be infectious, such as faeces and bodily fluids, decaying food, rats, and insects. This may have early led to a belief that such materials could be harmful and cause illness.

This might explain that even pre-modern humans – without knowledge of disease-causing microbes – may have deliberately attempted to spread infectious diseases and thus developed weapons that could be considered biological weapons. However, when we wish to study the use of biological weapons in pre-modern times, we encounter many problems. It is often difficult – and sometimes impossible – to ascertain with certainty which microbes were responsible for outbreaks of infection and epidemics in the past. We do not always know if what older accounts refer to as epidemics really involved infections. The descriptions of diseases are often imprecise and presented in a terminology that is foreign to modern medicine.

Furthermore, there are often quite uncertain and vague basis for the idea that possible infections were spread deliberately as biological warfare. Even in our own time, claims of the use of biological weapons are often part of the propaganda that inevitably arises in connection with war. The substance of such accusations is often impossible to assess.

Thus, there is reason to be sceptical when certain authors[6] claim that the use of biological weapons was widespread in antiquity and other pre-modern periods.[7] The evidence is often very thin and characterised by anecdotes. Nevertheless, there are several examples from ancient history of possible use of biological weapons that are of interest and often emphasized by historians.

Were the Hittites the First Ones?

The Hittites were an Indo-European people who, in the early 2000s BC, founded a kingdom in Anatolia in what is now central Turkey. The kingdom eventually became a great power that annexed several neighbouring states and came into conflict with another great power, Egypt. Around 1320 BC, the Hittites were at war with the neighbouring state of Arzawa in what is now western Turkey, emerging victorious from the conflict after two years. In the years preceding this war, the Middle East was ravaged by an epidemic that killed two Hittite kings. During the war, the Arzawans were severely impacted by this epidemic, which they believed the Hittites had inflicted upon them.[8]

On clay tablets inscribed in cuneiform found in the Hittite capital Hattusas (near the modern village of Boghazkoy), it can be read that the Hittite king desired that the enemy should be afflicted by the epidemic, and that the Hittites sent a sick ram and a woman across the border to Arzawa to achieve this.[6, 8]

Some authors have claimed that this epidemic was caused by the bacterium *Francisella tularensis*, which is responsible for tularaemia or rabbit fever as mentioned earlier. However, the documentation for this is weak.[7] The bacterium can infect a variety of animal species, including sheep, goats, and donkeys. Humans are infected from animals in various ways. As we have seen, tularaemia is still high on the list of possible biological weapons. The question is whether the Hittites have the dubious honour of being the first to use biological weapons. Although the Hittites made an active attempt to harm the Arzawans by inflicting an epidemic upon them – possibly tularaemia – there is little reason to believe that this was what led to the outbreak of the epidemic in Arzawa and to the outcome of the war. Tularaemia would likely have spread anyway.

It is also possible that the account of the woman and the ram being sent across the border may indeed describe a religious ritual. If so, it pertains to a ritual that was performed during epidemics, which we have other examples of in ancient Near Eastern history.[9] Such rituals were intended to cleanse the people of their sins, thereby leading to the cessation of the epidemic. The ram in the ritual is the basis for the term 'scapegoat'.

The Scythians' Poisoned Arrows

The Greek historian Herodotus from Halicarnassus in Asia Minor, who lived in the fifth century BC, is often referred to as 'the father of history'. Throughout much of his life, he travelled tirelessly through the countries around the eastern Mediterranean and wrote down everything he observed and heard about local conditions. While many historians tended previously to be quite sceptical of the truthfulness of Herodotus' colourful and often fantastic accounts, the current trend among historians is to take him more seriously.

On his many travels, Herodotus also visited the land of the Scythians. The Scythians were a nomadic people of Iranian origin who, around 700–800 BC, invaded what is now southern Russia, where they established a kingdom over wide areas north of the Black Sea.[10] They eventually came into contact with neighbouring peoples, including the Greeks, who both feared and were fascinated by this people, perceiving them as barbarians. Among the Scythians, both genders were tattooed. They were head hunters, as Scythian warriors were paid based on the number of heads of fallen foes they could display. Additionally – something that greatly shocked the Greeks – they drank unmixed wine! This last trait was unthinkable for a 'civilised' Greek.

The Scythians were feared warriors whom many perceived as invincible. They were experts in horse breeding and were excellent riders. They had developed an extremely effective bow made of layers of wood, horn, and sinew, with arrows equipped with special tips. The Scythians were superb archers, remarkably accurate even from horseback. Women also participated actively in warfare, which may have been the origin of the stories about Amazons in the ancient world.

Of special interest here is Herodotus's and other authors' accounts that the Scythians prepared their arrows in a special way to increase their lethal effect.[6, 10] This process was very complicated and was probably overseen by shamans. Decaying remains of venomous snakes were mixed with human blood, manure, and human faeces. The end product that the arrowheads were treated with undoubtedly contained masses of disease-causing bacteria. The arrows also emitted a terrible stench. It is reasonable to assume that the Scythian arrows would lead to severe infections in the arrow wounds. Additionally, knowledge of these poisoned arrows understandably instilled tremendous fear in the Scythians' opponents.

On Flowers and Bees

During their constant wars to expand their empire, the Roman Republic faced formidable opponents. One of these was King Mithradates VI of Pontus – a kingdom by the Black Sea. Mithradates, who reigned from 120 to 63 BC, was a dangerous and ruthless enemy, renowned and notorious for his knowledge of all kinds of toxins. He had also killed several of his family members with poison. The Romans had to fight three wars before he was finally defeated.

Initially, Mithradates had significant success, conducting a widespread massacre of Roman settlers in Asia Minor. Allegedly, 80,000 men, women, and children lost their lives in a single day. This naturally provoked the Romans. It was the famous Roman general Pompey – later Caesar's most dangerous rival – who ultimately defeated Mithradates.[11]

Mithradates not only fought with conventional weapons – including chariots equipped with sharp knives on the wheels that sliced enemy soldiers to pieces – he also used biological weapons. In his army, he had Scythian cavalry with their bows and poisoned arrows, which we have already encountered. Furthermore, he utilised his extensive knowledge of toxins to employ a sophisticated biological weapon, namely toxic honey.[11]

Flowers from some species of Rhododendron, which grow abundantly in Turkey along the Black Sea coast, contain the neurotoxin grayanotoxin.[11] When bees collect nectar from these flowers, the honey produced will contain this toxin. This had been known for centuries in these areas. In the third and final war against the Romans, Mithradates resorted to using this toxic honey. He ensured that pots of honey were placed where the Romans would find them. When the Roman soldiers indulged in the honey, they became intoxicated, collapsed, and some died. The soldiers thus became easy prey for Mithradates' warriors, who on one occasion killed 1,000 of them who had overindulged in the honey-sweet biological weapon.

It is worth noting that this honey, known in Turkish as 'deli bal' (meaning 'crazy honey'), is still produced – partly as a delicacy, partly because it is attributed with medicinal properties for several diseases.[6]

It is striking that biological weapons were not used by either the Greeks or the Romans in their countless wars and military conflicts. It has been suggested that this was due to a widespread belief that the use of such weapons was characteristic of 'barbarians', like the Scythians, and was unacceptable for 'civilised' Greek and Roman soldiers. It is interesting that one can also find ancient Indian texts emphasizing that this type of weapon is morally reprehensible. This indeed suggests that they were aware of forms of biological weapons. The moral condemnation of biological (and chemical) weapons is something we have seen throughout the centuries up to the present day.

Mongols and Catapults

In the mid-fourteenth century, the Black Death ravaged Europe, North Africa, and the Middle East, and likely the rest of Asia as well. This is one of the most horrific pandemics to have afflicted humanity. Most historians believe that the plague bacterium's point of origin for the pandemic in Europe was Crimea. There has also been a widespread belief that the start of the Black Death there was a result of the use of biological weapons – one of the most discussed examples of this in pre-modern times.[12]

The city of Kaffa – now Feodosia – in the Crimea was founded as early as 1255 by Genoese under an agreement with the Mongols of the Golden Horde, who were the rulers of southern Russia at that time. The relationship between the Genoese and the Mongols was often tense and could lead to war at various times. In 1346, Kaffa was besieged by a large Mongol army under Khan Janibeg. The siege had begun in 1343, with only one interruption. So far, the siege had been unsuccessful. The besiegers were now struck by the plague, which had likely come from Central Asia via the Silk Road – the ancient trade route between east and west.

Thousands of the Mongol soldiers died from the plague, and Janibeg decided to lift the siege. However, he resolved to launch a final attack as a parting gift to the besieged Genoese. Janibeg commanded his men to use siege machines – *catapults* – to hurl plague-ridden corpses in large quantities over the city walls into Kaffa. The Khan's intention was undoubtedly to harm the besieged. But how did he intend to achieve this?

The story that is repeatedly referred to actually has only one source – the Italian Gabriele de' Mussi, who wrote an account of the siege of Kaffa. He was a lawyer in Piacenza, some distance north of Genoa. It is uncertain whether de' Mussi was present at the siege, but he likely spoke with eyewitnesses from there. He recounts that 'mountains of dead were thrown into the city' and that the Khan believed 'the unbearable stench would kill everyone inside'. The result was that the deadly plague then broke out among the besieged. Was this a result of the Khan's use of plague corpses as a biological weapon? Or had the plague infiltrated the city with infected rats, which were the most common transmission mechanism? In the latter case, fleas transmit the plague bacterium from rat to human.

Gabriele de' Mussi is regarded by historians as a relatively reliable source regarding the main features of the account.[7] However, several more recent historians doubt that it was the plague corpses that brought the infection into Kaffa. Sceptics argue, among other things, that plague-bearing fleas quickly abandon dead and cold bodies to find new living hosts – and therefore could hardly have gone along on the catapult journey into Kaffa. These historians thus lean towards the theory that it was rats that brought the plague bacterium into the city.

However, those who argue that the Khan's use of plague corpses as biological weapons caused the outbreak of plague in the city also have good arguments. They assert that the besiegers' lines were probably 200–300 metres from the city walls. Since rats are very sedentary and do not move many metres from their nests, it is believed that plague-infested rats from the Mongol camp were unlikely to have wandered into the city. Moreover, the defenders of the city could also have become infected by direct contact with the bodies when these were removed in large numbers.[12]

Overall, de' Mussi's account of the Khan's attack with plague corpses is quite plausible. If so, it was the use of biological weapons that initiated the ravages of the Black Death in Europe. Some Genoese chose to flee from the plague-ridden city via the sea route, which was still open. But the galleys carried with them both infected people, rats, and fleas. After just a few days at sea, the first plague deaths occurred. The Genoese galleys sailed from the Black Sea through the Bosporus and brought the plague to Constantinople, which was hard hit. Furthermore, the Genoese carried the plague to several port cities in both the eastern and western Mediterranean. One after the other, the rest of Europe was hit – including Norway in 1348.[13]

It is likely that it was not only the Genoese galleys from Kaffa that brought the plague into Europe. The plague bacterium likely had other points of entry, both via ships from other ports and along caravan routes from the East.

In any case, Gabriele de' Mussi's dramatic account of the siege of Kaffa will remain in the history of biological weapons – also as an important reminder for us today that the use of biological weapons can lead to extensive epidemics.

There are examples of the history of Kaffa repeating itself in later centuries, albeit on a smaller scale. During the long-running war in the Baltic between the Swedes under Charles XII and the Russians under Peter the Great, the plague also ravaged. After the Russians had won their decisive victory in the Battle of Poltava in Ukraine in 1709, they besieged the Swedes in the city of Reval in 1710. There they dumped a number of corpses – undoubtedly also plague corpses – into a river that flowed through the city. Three-quarters of the city's 20,000 inhabitants died from the plague – but unlikely because of the Russians' alleged biological weapon attack.[14]

From Leprosy to Plague and Rabies

Concepts of infection regarding certain diseases have, as mentioned, existed for centuries alongside other explanatory models. This has evidently inspired the belief that potentially infectious materials can be used as weapons. Ignorance about microbes and their actual modes of transmission often rendered such early attempts at using biological weapons unsuccessful. There are several examples of this.

In the late 1400s, Italy was the scene of constant hostilities between Spain and France in shifting alliances with Italian city-states and princes. During the struggle for Naples in 1495, it is reported that the Spaniards used wine mixed with blood from patients suffering from leprosy as a weapon against the French.[7] Since from biblical times, leprosy patients had been perceived as unclean – infectious – and were strictly isolated from society. Today we know that this infectious disease does not spread through food and drink but likely through nasal secretions and only with prolonged and close contact.[13] If the Spaniards truly attempted to use leprosy-contaminated wine as a biological weapon in 1495, this would have been ineffective. The sources of this account are somewhat unreliable.

In contrast, it should be noted that the new dreadful disease syphilis appeared in Europe at the same time and severely affected the French during their retreat from Naples. However, this was not a result of biological weapon use but rather of the French's interactions with Naples' prostitutes, who were likely infected by Spanish soldiers.[13]

In 1650, the Polish artillery general Jan Kazimierz Siemienowicz published a work in Latin on the art of artillery – *Artis Magnae Artilleriae Pars Prima* (*The Great Artillery Art. First Part*). In this, he describes how to coat artillery balls with saliva from rabid dogs to increase their destructive effect. Some authors claim that such balls were also used in a battle, but this is uncertain. In any case, rabies would not have resulted from using such ammunition. Rabies is caused by a virus that is only transmitted by the bite of an infected animal. The general's biological weapon would have been ineffective.[7]

Several authors report that Louis XIV of France was offered a supposed biological weapon by an Italian inventor, although details are not known.[7] The king declined and granted the inventor a pension on the condition that he never revealed the secret. If this story is true, it suggests that Louis had a reluctance to use such weapons, a stance that has been common throughout history.

Venice was a maritime superpower in the eastern Mediterranean for centuries and had several strongly fortified possessions there. In a long series of wars, the Venetians fought against another great power in the region – the Ottoman Empire. There were both trade-political and religious reasons for this. Venice was the foremost protector of Christianity against Islam in these areas.

During one of these wars, an Ottoman army besieged the city of Candia – today's Heraklion – which was the capital of the Venetians' possessions in Crete and central to their dominance in the eastern Mediterranean. This siege lasted from 1648 to 1669 – the longest in military history. It was crucial for the Venetians to break the siege.

In 1649, a high-ranking Venetian leader, Lunardo Foscolo, proposed that the siege be ended by exposing the Ottoman besiegers to a dangerous liquid that he called the 'quintessence of plague'. Plague was well known in Crete, where, since the Black Death, there had been 20 epidemics.

Of course, Foscolo did not want the Venetians' use of plague as a weapon to become known. He was aware that the project was morally questionable and in violation of the rules of war. Therefore, he suggested that similar plague attacks should also be directed at all the other Ottoman camps around Crete, so that the desired outbreak of plague would appear as a natural epidemic.

Foscolo had recruited a Venetian doctor to prepare the plague weapon from fluids pressed from spleens, buboes, and wounds of patients who had died of plague during an ongoing epidemic in Dalmatia. They planned to impregnate textiles with the fluid and then distribute the materials in such a way that the Ottoman soldiers came into contact with them and became infected.

The planned plague attack was, however, postponed from the autumn of 1649 to 1650 because it was believed that plague would spread more effectively in the summer months. The responsible physician was imprisoned for safety over the winter, probably to prevent leaks about the plan, but then it seems the Venetian leaders lost interest in the whole project. The reason may have been that the Venetians achieved a series of military victories and thus assumed that the siege of Candia would cease. However, this did not happen, and the siege lasted until 1669.

It is quite doubtful whether Foscolo's biological weapon would have been successful. How long *Yersinia pestis* – the plague bacterium – survives outside the host organism depends entirely on the environment. Even though it can remain alive in soil for many weeks, it is unlikely that it would have survived for months in Foscolo's plague fluid.

'The Most Terrible of All the Ministers of Death' – Smallpox

With full justification, one of England's greatest historians of the nineteenth century – Thomas Babington Macaulay – with Victorian pathos described smallpox as 'the most terrible of all the ministers of death'. Over time, this virus has claimed millions of lives. In the twentieth century alone, there were 300 million deaths due to smallpox – three times as many as in all the wars of the century. Smallpox has also been used as a biological weapon throughout history.

When members of the human species Homo sapiens began migrating out of Africa between 60,000 and 100,000 years ago, they spread to different continents. They encountered new microbes that varied significantly from continent to continent. This affected human genes. Through the evolutionary process, more or less subtle differences in the genome – the hereditary material – arose from region to region. The result was often a certain increase in genetically determined resistance to the microbes found in each region. Additionally, individuals developed more or less effective immunity after exposure to these microbes. Therefore, resistance to many microbes varies significantly across different parts of the world.

These conditions had dramatic consequences during what I have termed the second major transition in human infection history, after the first one when humans settled permanently and began farming and animal husbandry.[13] The second transition occurred when Europeans, starting in the late fifteenth century, discovered new continents – first America and then Australia. The indigenous peoples of these continents had been in contact with an entirely different microbial flora than the Europeans and thus had neither any genetic defence nor immunity against many microbes to which Europeans had far better resistance.

Biological Weapons Against Indigenous Americans

An important example of differences in resistance to microbes between continents is the smallpox virus. In Europe, people had been exposed to this virus since childhood for centuries. Those who did not die at the first encounter developed lifelong immunity. On the American continent, however, it had fatal consequences when the indigenous population came into contact with the smallpox virus brought by Europeans. Widespread epidemics spread over large areas, accompanied by horrifying mortality, resulting in significant demographic consequences.

This first occurred in Mexico when Spanish conquistadors led by Hernan Cortez invaded in 1519. It is overwhelmingly likely that the Spaniards would not have succeeded in crushing the powerful Aztec Empire if the smallpox virus had not been their ally and struck the Aztecs.[13,16] In the same way, the smallpox virus paved the way for Francisco Pizarro and his men when they conquered the Inca Empire in Peru about ten years later.[13]

In the following centuries, the tragedy repeated itself over and over as the remaining indigenous populations in North and South America came into contact with the smallpox virus introduced by Europeans.

Overall, it may not be surprising that the idea of using the smallpox virus as a biological weapon against the indigenous population arose, but there is no evidence that the Spanish conquistadors of the 1500s thought along these lines.[7] On the other hand, we know for certain that the smallpox virus was used against the Indians in North America toward the end of the Seven Years' War from 1756 to 1763 when England and France fought against each other, including in North America. England emerged victorious from the war and captured the French possessions around the Great Lakes. The Indians in the area were very dissatisfied with their new masters. This led to the so-called Pontiac Rebellion in 1763, in which a massive alliance of various tribes led by Ottawa chief Pontiac waged war against the English. The Indians captured several forts and killed or captured hundreds of English soldiers and colonists.[17]

An English garrison at Fort Pitt, located at present-day Pittsburgh, was besieged by Delaware Indians during the war. At the end of a negotiation meeting between the English and the besiegers, an English officer, Captain Simeon Ecuyer, gave the Indians wool blankets and a handkerchief from the fort's hospital where there were some smallpox patients at the time. The intention was to infect the Indians with smallpox. The captain wrote in his diary, 'I hope it will have the desired effect.'

We know for certain that the idea of using smallpox against the Indians was not just a single officer's concept. Letters a few weeks later from the English commander Sir Jeffrey Amherst show that he both approved and actually recommended the use of this strategy against hostile Indians. Amherst later became a baron and ended his days as a field marshal.

This is the best-documented example of the use of biological weapons before the modern understanding of infection based on microbes emerged in the second half of the nineteenth century. Again, we see that even in earlier times, there was a practical understanding of some form of contagion.

Many have later condemned General Amherst's desire to use the smallpox virus as a biological weapon. He was likely aware that this violated the rules of war in general, but he evidently believed that the war against the Indians was so irregular – including murders and scalping by the Indians – that it justified the strategy. In a letter to a subordinate officer, he wrote: 'You should try to infect the Indians with wool blankets, in addition to trying every other method that might contribute to exterminating this terrible race.'

Was this strategy effective? It is uncertain. The smallpox virus primarily spreads through droplet infection, that is, through relatively close contact. However, the virus can also survive in dried form for longer periods on textiles, such as clothing, wool blankets, and bed linens, and can result in infection through contact and aerosols. Captain Ecuyer's wool blankets may have led to infection. Nevertheless, it was only a relatively small number of Indians who died, and this did not lead to the lifting of the siege. Furthermore, the smallpox virus was circulating quite freely in these areas already.[17]

There is also another example of the use of the smallpox virus as a biological weapon against hostile Indians. This is admittedly not as well documented as the one mentioned but is still quite likely.

Traffic along the trade route from Saint Louis to Santa Fe suffered significant losses of life and property around 1830 due to constant attacks from Plains Indians, particularly the Pawnee tribe. In 1831, some traders chose to spread garments and tobacco impregnated with smallpox material as gifts to the Indians they met along the way. Some of these gifts ended up with a Pawnee tribe on a warpath. The Indians were struck by a smallpox epidemic, and within a few days of returning home, the virus exterminated half the tribe.[17]

There are no other documented examples of the smallpox virus being used as a biological weapon against Indians in North America. However, on several occasions, rumours of such practices arose, particularly among the native population, where naturally occurring smallpox epidemics occurred frequently.

In South America, which was colonised by the Portuguese and Spanish, conflicts with the indigenous population naturally occurred. The famous French social anthropologist Claude Lévi-Strauss claims that the Portuguese deliberately attempted to infect the Indians in the Amazon area with the smallpox virus as early as the 1500s.[7] In his partially autobiographical book *Tristes Tropiques* he states that until relatively recently it was a 'favourite pastime' for Brazilian men 'to collect infected garments from smallpox patients in hospitals and hang them up together with other gifts along the paths where the tribes still roamed'.[18] Lévi-Strauss regarded these reports as reliable. There are no reports of similar practices in the Spanish possessions in South America.[7]

Smallpox Against Aborigines

The indigenous population of Australia – the Aborigines – had also been isolated from other continents for tens of thousands of years and had therefore been shielded from a number of microbes that did not exist in Australia, including the smallpox virus. This would have fatal consequences just as we saw on the American continent.

In January 1788, the first English colonists, including 750 convicts, arrived at Botany Bay near present-day Sydney. In addition to many other difficulties, they faced increasing hostility from the local Aborigines.

In April 1789, the colonists learned that smallpox had broken out among the Aborigines in the area.[19] Bodies were constantly being found on the beaches and floating in the sea. Smallpox spread like wildfire among the indigenous population of south-east Australia, with fleeing Aborigines from the Botany Bay area. Several modern historians believe that this made it easier for the English to colonise Australia.

What was the background for this tragic smallpox epidemic among the Aborigines? Several theories have been discussed.[7] Some believe that the virus had recently arrived on the Australian continent via fishermen from Sulawesi (formerly called Celebes) in Indonesia, where smallpox existed. This theory assumes that after its arrival in the north, the virus subsequently spread among the Aborigines throughout the continent down to the south coast. This is unlikely, as the Aboriginal tribes were very dispersed, making virus transmission difficult.

There is no evidence that the English colonists brought the smallpox virus naturally; there were no cases of smallpox among the colonists.

The theory that probably has the most supporters today is that the smallpox virus was used as a biological weapon by the English against the Aborigines in the increasingly threatening conflict.[7] We know that an English naval doctor had with him a bottle containing smallpox material that was intended for what was called variolation, that is, the first form of vaccination against smallpox that was implemented in Europe from the mid-eighteenth century. In variolation, material from lesions of smallpox patients, which contained live viruses, was scratched into the skin. Was this material, in some unknown way, used to trigger smallpox among the Aborigines?

Proponents of the biological weapon theory emphasize that at least one of the naval officers had served previously under Sir Jeffrey Amherst during the Pontiac Rebellion and thus may have been familiar with the use of the smallpox virus as a biological weapon at that time.

There is no definitive evidence that the Aborigines were intentionally exposed to the smallpox virus, but the indications are quite strong. The alternative theories are less likely.

We do not know for sure what the population of Aborigines was before the arrival of Europeans, but it is estimated to be in the hundreds of thousands. Sixty years after the arrival of the English in 1788, the Indigenous population had decreased by 60–90%. Smallpox likely bore significant responsibility for this, but other imported infections such as measles, tuberculosis, and sexually transmitted diseases also contributed.

The infection history of the Aborigines is only one of many examples of the tragic consequences when a potentially dangerous microbe is introduced into a population that has never been in contact with it and therefore lacks any immunity. It also serves as a possible warning of the dreadful consequences that the use of biological weapons can entail.

Biological Warfare During the American Revolutionary War

The quotes from Amherst's aforementioned letter could suggest that the English, during the Pontiac Rebellion, justified the use of smallpox as a weapon against their Indian adversaries due to their barbaric warfare, and possibly also from racist considerations. However, events during the fighting between the English and the 13 colonies during the American Revolutionary War from 1775–83 suggest that this is not the whole explanation.

The English troops were largely immune to smallpox – partly due to previous infection and partly due to so-called variolation. The American colonial army, on the other hand, had generally low immunity to the smallpox virus. Variolation was forbidden. Smallpox consequently ravaged among the American soldiers and claimed more lives than combat actions. On several occasions, this had serious consequences for the military efforts of the colonial army.[13]

The English exploited this in several ways. When they were besieged by the Americans in Boston in 1775, the English variolated American residents and sent them out to the besiegers. Since those variolated were contagious for some time after variolation, they could spread the virus within the American colonial army. This form of biological warfare was possibly repeated several times by the English.

These problems caused George Washington, who led the American colonial army, to introduce variolation for the army. This significantly reduced the smallpox issues.

During the fighting in the southern states, where the large slave plantations were located, it often happened that slaves escaped and sought refuge in the English army. Here, they were quickly infected with the smallpox virus. The English sent the infected slaves back to the plantations to spread the virus.

Were Bioweapons Used During the American Civil War?

During the bloody American Civil War, it is estimated that 660,000 soldiers lost their lives. A full two-thirds of these deaths were due to infectious diseases rather than combat actions. On several occasions, officers from the Confederate states believed that the Union states attempted to use the smallpox virus as a bioweapon against them. Rumours also circulated that the Union forces had vaccinated prisoners of war with a smallpox vaccine contaminated with syphilis material to spread syphilis. There is no evidence supporting these claims. The huge burden of infection among both warring parties was likely related to the inevitable connection between war and infectious diseases – a connection that has been recognised for time immemorial.[1,7]

Chapter 3

The Discovery of Bacteria – the Bacteriological Revolution as a New Platform for Bioweapons

In the first half of the nineteenth century, most people believed that many common infectious diseases, such as syphilis, gonorrhoea, smallpox, and measles, were contagious from person to person. However, this was not associated with any living pathogen. Scientists likely were aware of the existence of bacteria that the Dutchman Antonie van Leeuwenhoek had discovered in the late seventeenth century. Nevertheless, no one had convincingly documented that such microorganisms could cause disease in humans. However, several researchers began to assert this, which became the foundation for modern theories of contagion.[13]

Eventually, a significant polarisation arose between supporters of the new contagion theories and those who still adhered to the miasma theory, albeit often in modified forms. Many now believed that disease-causing 'miasma substances' did not only act through the air, as originally thought, but could also be transmitted via clothing and objects.

Then came the major breakthrough –what we call the bacteriological revolution – marked by a clear demonstration that many diseases were caused by infections with bacteria that could be transmitted from person to person in various ways. Of the many pioneers in this field, Robert Koch in Germany and Louis Pasteur in France were the foremost. In 1876, Koch published his evidence that the disease anthrax was caused by a bacterium. In 1882 he discovered the bacterium that causes tuberculosis and in 1883 the bacterium that causes cholera. Concurrently, Pasteur showed that other diseases in plants, humans, and animals were caused by various bacteria. During these years, the bacterial causes of numerous other infectious diseases were identified – some by Koch's and Pasteur's students, and also by other researchers.[13]

Even Pasteur discovered that certain pathogens were far smaller than bacteria and could not be seen with ordinary light microscopes – what we now know to be viruses. However, it was not until 1930 that viruses could be observed using the newly developed electron microscope.

These groundbreaking discoveries truly deserve the designation 'revolution'. They overturned over two thousand years of misconceptions about infectious diseases and also had implications in areas far beyond infectious medicine.

However, the bacteriological revolution also had a darker side: it laid the groundwork for a far more rational development of bioweapons that were considerably more effective than previous, often completely ineffective attempts.

In the early years following the breakthrough of the bacteriological revolution, the vast majority of scientists were focused on the medical advancements it would bring. The idea of using microbes as bioweapons was likely distant from the minds of researchers like Koch and Pasteur.

By the end of the nineteenth century, assassinations of politicians, state leaders, and royalty were far from uncommon. It was primarily various groups of anarchists who were responsible for this. The anarchists quickly began to use Alfred Nobel's new invention, dynamite. Although it turned out that dynamite was often more difficult for terrorists to handle than more conventional weapons like pistols and daggers, the anarchists believed that the psychological effect of dynamite attacks was important as part of what they called 'the propaganda of the deed'.[21] Just as the anarchists early on embraced dynamite, some anarchist theorists also expressed interest in using bioweapons for assassinations.[7] For unknown reasons, this did not have any consequences for the anarchists' practices. However, the fear of this occurring was expressed in some parts of the popular press.

There are many examples of science fiction writers predicting future breakthroughs in technology and research. We also see examples of this concerning the idea of bioweapons, which appears early in science fiction literature.

An early example of this is the English author H.G. Wells' short story 'The Stolen Bacillus' from 1895, where cholera bacteria are imagined to be used as a bioweapon by anarchists.[22] At this time, the English population feared anarchist attacks, which were frequent on the Continent. Wells was clearly inspired by the dire consequences of cholera epidemics that had ravaged

London a few years earlier.[13] Robert Koch had also recently discovered the cholera bacterium.

In the story, a bacteriologist shows a test tube containing cholera bacteria to a mysterious visitor, detailing the deadly consequences if the contents were poured into London's water supply: '... mystical and traceless death, swift and terrible death, painful and undignified death'. The visitor is impressed and exclaims: 'These anarchists – scoundrels – are idiots, blind idiots – to use bombs when something like this is available!' He seizes the opportunity to steal the test tube, planning to pour its contents into London's water supply to trigger an epidemic. He escapes in a horse-drawn cab, closely pursued by the bacteriologist who discovered the theft. By accident, the test tube is smashed. The thief drinks the contents while shouting 'Vive l'anarchie!'. During his ongoing flight, he rubs against as many people as possible to spread the infection. This is an extraordinarily ineffective way to spread cholera! Moreover, it turns out that the stolen test tube did not contain cholera bacteria, only a completely harmless bacterium. Regardless, Wells' story highlights the possibilities bioweapons can present for terrorists.

The prolific author Jack London was also fascinated by bioweapons. In the short story 'Yah! Yah! Yah!' from 1909, a European punitive expedition in the Pacific intentionally exposes the Polynesian population to the measles virus, leading to numerous deaths. London exploits the knowledge of the historically tragic results when the measles virus has been introduced into isolated populations that have not previously been exposed to the virus, thereby lacking immunity against it.

London also addressed the topic of bioweapons in the novel *The Unparalleled Invasion*, in which Western nations attack China with a variety of different bioweapons dropped from airships.[23] The entire population of China is exterminated. The theme of this book reflects the widespread fear in the USA at the time of the 'Yellow Peril', represented by Asians threatening the dominance of the white race. The action also illustrates the heroisation of scientists that characterised the period.

During the Second Boer War from 1899 to 1902, fought in South Africa between the Boers and the English, rumours circulated that the Boers were using bioweapons. It was claimed that they used the bacteria causing anthrax and glanders against the English's horses, but this has never been documented.

Chapter 4

When the Great War Raged – Bioweapons During World War I

When World War I broke out in 1914, several decades had passed since the scientific breakthroughs of the bacteriological revolution. The bacteria causing a wide range of common infectious diseases had been discovered, and the important modes of transmission were clarified. Although viruses could not yet be seen in the microscopes of the time, their existence as a cause of many infectious diseases was recognised. Vaccines against several microbes had already been developed. It was essentially inevitable that the possibility of using microbes as weapons would eventually arise.

Germany and France had led the way in the bacteriological revolution, bearing proud memories of Robert Koch and Louis Pasteur, and both countries had very strong microbiological research environments. This may explain why Germany, and to a certain extent France, were the first countries to genuinely invest in the military use of bioweapons during World War I.

The First Comprehensive Bioweapon Programme

Germany built an extensive military programme for the development and use of bioweapons from 1915 to 1918.[24] This was completely controlled by the German Army's General Staff, without any political oversight. Several hundred officers worked in the bioweapon programme, training personnel and planning and executing operations. They undoubtedly collaborated with microbiological research environments.

It is remarkable that the German bioweapon programme was established despite the General Staff's guidelines stating that biological warfare was illegal. The explanation is likely that they believed the prohibition only applied to the use of bioweapons against humans, not against animals. It is a well-known fact that Germany used its bioweapons exclusively against animals during World War I, as these were considered vital for the enemy's war efforts. On several occasions, proposals were made to the General Staff to also attack humans with bioweapons, but this was categorically dismissed as unethical.

In essence, it was horses and mules that were the targets of Germany's biological warfare. During World War I, the warring armies totally depended on horses for the transportation of equipment and supplies. There was significant loss of horses during the fighting, and they had to be constantly replaced. Several neutral countries in the early years of the war were important suppliers of draft animals to the Allied countries, including Spain, Argentina, Romania, and the USA, which remained neutral until 1917. For other reasons, Finland and Norway were also drawn in.

In its bioweapon programme, Germany focused on the bacteria that cause glanders (*Burkholderia mallei*) and anthrax (*Bacillus anthracis*). Both of these can cause serious illness in horses and mules. These bioweapons were transported to various countries, among other ways, by German diplomats, sometimes using submarines. In the USA and Spain, the relevant bacteria were also produced in secret, local laboratories. The deployment of the bacteria to the animals that were the targets occurred in various ways, often carried out by pure amateurs. For example, in the USA, dock workers were employed for this purpose.

There were also some examples of Germans using bioweapons against enemy states. In France, German agents attempted to infect cavalry horses with glanders.

Was Germany's extensive and resource-intensive bioweapon programme actually effective? This is uncertain. German documents that could provide information about this have disappeared. Much suggests that those responsible for the programme believed it had yielded results, but later investigations provide no evidence to support this. In the USA, there was a fairly widespread belief that German sabotage was the cause of several outbreaks of anthrax in horses across the country – but this could also have been natural epidemics.

Regardless, Germany's bioweapon programme during World War I was historically significant in several ways. It was the first state programme for the use of biological weapons in war. It was also the first bioweapon programme based on scientific facts from microbiology. Furthermore, Germany's is one of only two documented examples of the use of bioweapons in war in modern times. The other was by Japan, which we will soon examine.

France's Modest Bioweapon Investment

France also sought to use bioweapons during World War I, but the French programme was far less extensive than the German one.[25] The French also aimed to target animals – not humans. Germans claimed that French agents attacked horses in Switzerland, which were to be sent to Germany, with glanders bacteria.

Additionally, France attempted to utilise French prisoners of war in Germany as bioweapons agents. On at least two occasions, the Germans found bacterial cultures with prisoners of war that were evidently intended for infecting cattle and pigs.

The Baron in Karasjok

A particular episode of Germany's biological warfare during World War I took place in Norway. Swedish Baron Otto von Rosen, an agent for Germany, operated on Norwegian soil with plans for using bioweapons.[26] He was an officer and a highly skilled marksman, and in 1908, he had participated in the Summer Olympics in London.

Von Rosen had strong German sympathies and was recruited in 1916 by the German General Staff to conduct biological warfare against Russia. At that time, Finland was a grand duchy, part of the Russian Empire. Von Rosen connected with the Finnish liberation movement. He attacked horses and cattle near Russian military units with anthrax bacteria in small glass tubes hidden in sugar lumps that he placed in the animals' feed. He then moved his operations to Norway.

In January 1917, he was arrested along with two Finnish collaborators as suspicious individuals in Karasjok. His luggage contained explosives, maps of Russia and Finland, and the aforementioned sugar lumps with glass tubes that, upon later examination, turned out to contain anthrax bacteria. Additionally, a small bottle containing typhus bacteria was found.

Von Rosen was sent to Kristiania for further interrogation. He declared that he was working for Finland's independence with support from Germany. After three weeks in custody, he was allowed to leave the country. This was before everything about the contents of his luggage was known.

Later, it was realised that the baron's target for the bioweapons was horses – possibly also reindeer – that were to be used as draft animals in Finnmark for transporting weapons from England to Russia, which was an ally.

Some of the sugar lumps from von Rosen's luggage ended up at the Police Museum in Trondheim. They were rediscovered around 80 years later, and the contents were analysed at a special laboratory in Porton Down, England. There, DNA from the anthrax bacterium was detected.

Chapter 5

Towards a New War – Bioweapons Between the Wars and During World War II

In the years following World War I, interest in bioweapons grew in several nations. It was no longer just animals that were the targets, as in the earlier German bioweapons programme. Now, it was primarily humans that were targeted – both soldiers and civilian populations in enemy countries. Japan and France led the way in bioweapon development. However, unlike France, Japan used such weapons in war, and the Japanese displayed a ruthlessness in testing their weapons that had seldom been seen before or since.

Horrors in the Land of the Rising Sun

By the beginning of World War I, Japan was on its way to becoming a great power. After its victory over Russia in the Russo-Japanese War from 1904 to 1905, the country had pursued an aggressive policy toward China, a nation that was, in many ways, disintegrating following the fall of the imperial dynasty in 1911. Japan had secured a foothold in Manchuria, which was technically part of the Chinese Empire.

In Japanese society, the military had a significant influence on political life. The army was directly subordinate to the emperor. It was the government that ruled, but the military had considerable influence on the government, which had to take significant military interests into account. The military leaders and the politicians who supported them had grand imperialist ambitions that were initially aimed at China and Manchuria. Military rearmament was a high priority. This was the backdrop for the Japanese bioweapon programme, which would become one of the most extensive the world had ever seen.[27]

A key figure in developing the Japanese bioweapon programme was Ishii Shiro, a young military doctor with the rank of lieutenant in the mid-1920s. After reading an article about the Geneva Protocol of 1925, which prohibited the use of bioweapons, he had an idea. If other countries refrained from developing and using such weapons, the Japanese military should focus precisely on these weapons. This would give Japan an advantage over any enemy.

Ishii was a unique personality. He was intelligent, arrogant, highly ambitious, and had the ability to impress his superiors. He had fanatical followers but also opponents. Ishii quickly advanced in his career. He invented a new system for filtering water during epidemics. His confidence was evident when he demonstrated his invention to the emperor by urinating into the filter and inviting the emperor to drink the result. The emperor politely declined.

After a two-year study trip, during which he visited microbiological research laboratories in Europe, the Soviet Union, and the United States, he was appointed professor of immunology at the army medical faculty in Tokyo. There, he built a dedicated research laboratory where he began research on bioweapons. It is clear that he had already planned to ignore common ethical research considerations from the outset. He quickly realised that he could not conduct his projects in Japan due to potential protests if his activities became known. Therefore, he wanted to continue his research abroad, and the opportunity arose when he was permitted to establish himself in Manchuria in 1932.

The Death Factory in Manchuria

In Manchuria, circumstances were ideal for Ishii's sinister enterprise.[27] In 1931, the Japanese army stationed in the country to protect Japanese citizens had taken control of the land in a coup. Manchuria had now become a Japanese puppet state named Manchukuo, governed by the army and the brutal military police, the Kempeitai, which could effectively shield Ishii's research activities from critical eyes. There were several oppressed ethnic groups in the country that could be used for the bioweapon experiments Ishii wanted to conduct.

He initially settled in the large city of Harbin, where he gained access to prisoners from the local jail. However, he feared that he could not avoid publicity even here and moved shortly afterward to Zhong Ma, about 100 miles south of Harbin. There, he built a large complex with multiple buildings surrounded by a high wall with an electric fence. The facility could house 500–1,000 prisoners. Here, Ishii seriously began his experiments on humans, infecting them with bacteria that cause plague, anthrax, and glanders. In several experiments, infected subjects were dissected alive without any form of anaesthesia, which Ishii believed could interfere with the results.[27, 28]

After two years of numerous experiments of this nature, causing unimaginable suffering for the victims, some prisoners managed to escape. The secret about the horrors in the Zhong Ma camp leaked out. Shortly afterward, former prisoners managed to blow up the entire complex. Ishii's superiors were then willing to let him build an even larger facility for bioweapon research. This new facility was constructed at Ping Fan, a few miles outside of Harbin, and was enormous – 6 square kilometres in size. It consisted of 150 buildings, including laboratories, two prison buildings, barracks for staff, three crematoria, separate buildings for autopsies and dissections, storage buildings, animal stables, and its own power station. There were recreational facilities for staff, including a library, swimming pool, and brothel. It also had its own farm for growing vegetables and fruits. There was even an airport with Ishii's private fleet of aeroplanes. A dedicated railway line ran from Harbin to the facility and was used to transport prisoners for experiments, often at night.

At the Ping Fan facility, there were 3,000 staff – scientists, technicians, and soldiers. From 1941, this facility was designated Unit 731. What was, in reality, a death factory became the most extensive bioweapons facility in history and the very symbol of Japan's bioweapons programme.[27, 28]

Men, women, and children were used in experiments at Unit 731. They included criminals, war prisoners, communist partisans, Russian emigrants, political opponents of the regime, mentally ill individuals, and a random selection from the local Chinese population. The activities at Unit 731 were top secret and shielded by the military and the Kempeitai. All correspondence from the Japanese staff was censored, and strict travel restrictions were in place. The official version of the operations was that the facility was a lumber mill. With rather bad taste, Ishii and his colleagues referred to the prisoners as 'maruta' – logs.

In the laboratories, vast quantities of the bacteria intended for the experiments were produced. For example, 1 ton of cholera bacteria, 900 kilograms of typhoid bacteria, 600 kilograms of anthrax bacteria, and 300 kilograms of plague bacteria were produced annually.

The experiments the prisoners underwent aimed to develop the most effective bioweapons possible, and a range of pathogenic bacteria were studied. Researchers investigated what dosages of microbes were most effective and which modes of transmission were best. The prisoners were also used extensively to test new vaccines against common infectious diseases.

Numerous experiments also focused on developing the most effective means of spreading microbes as bioweapons. Artillery shells, various bombs, and spray systems from aeroplanes were tested in a dedicated test area that Unit 731 controlled north of Harbin. The bombs proved most promising. In some of these experiments, subjects tied to stakes were subjected to shrapnel from aerial bombs and grenades containing microbes, to assess the possibility of causing fatal wound infections.

No one survived these experiments. Either the victims died during the trials, or they were subjected to vivisection – dissected alive – or put to death in various ways after a time when the experiment was deemed completed. All were autopsied, and the results were archived. Remains from the victims in the form of tissue samples were also stored.

The exact number of subjects who lost their lives at Unit 731 and the many satellite facilities that were established around Manchuria and China will likely never be known. It undoubtedly involved thousands – perhaps more than 10,000.

There was also intense research on developing bioweapons against grain types and vegetables, particularly those species relevant to that part of the world.

Unit 731 is the most discussed centre of Japanese bioweapon research, but it was by no means the only one. Another facility with high activity was established at a military medical institute in the city of Mukden. This was led by Kitano Masaji, a military doctor with the same background as Ishii – also with a professorship. His colleagues considered him a more skilled researcher than Ishii, and he was equally ruthless in his human research. He personally led many experiments. After ten years at the centre in Mukden, where he was known as 'The Emperor', he took over command of Unit 731 when Ishii was transferred.[27]

A third complex for biomedical research was built near the city of Changchun, which was the capital of Manchukuo.[27] The facility was named Unit 100 and was led by veterinarian Wakamatsu Jujiro. This top-secret complex was nearly as extensive as Unit 731 and was built in a similar manner. Its main focus was primarily the development of bioweapons against animals and plants. However, significant human experiments were also conducted there. There was some cooperation with Ishii and his group. Large quantities of bacteria were also produced here for use in experiments and in the field.

In addition to these larger centres, a network of smaller satellite centres for bioweapon research was gradually established in Manchuria, China, French Indochina, Thailand, Burma, and possibly also in the Philippines. Beyond the exceedingly high, albeit uncertain death toll from operations at Unit 731, there are also the victims who lost their lives at all the other 'death factories' involved in Japan's gruesome bioweapon programme. The true number remains unknown.

Who were the scientists driving this operation? It is both surprising and shocking that it was not just a handful of eccentric personalities devoid of any empathy, possibly even with sadistic traits. It is a fact that the brightest minds from Japanese universities were lining up to work under Ishii in the bioweapon programme – fully aware that this involved experimenting on humans. A colleague of Ishii later stated: 'Most microbiologists in Japan were in some way connected to Ishii's work. He mobilised the majority of universities in Japan as contributors to the research at his centre.' The results of the gruesome experiments on human subjects were often published in reputable scientific journals, albeit misrepresented as experiments involving various primate species.[27]

The motives of these scientists varied. Some were fanatical nationalists who believed that anything that strengthened Japan militarily was paramount over all other considerations. Many were lured by promises of good pay, career opportunities, and interesting, new research without constraining ethical limitations. Strongly racist attitudes where races other than Japanese were considered inferior likely influenced their views on experimenting on humans from such groups. Japanese soldiers were also fanatical supporters of the samurai's ancient bushido code, which was highly esteemed in Japanese society. This bushido attitude fostered deep contempt for soldiers who surrendered. They were seen as dishonourable and could be treated without compassion. This made it easier to experiment on prisoners of war.

It is highly plausible that Emperor Hirohito, who had an education in biology, was informed about the Japanese bioweapon programme. It is also a fact that the emperor's younger brother, Prince Mikasa, himself visited Unit 731.

Japan's Biological Warfare

Eventually, Ishii became eager to test his bioweapons in practice in the field.[27] The opportunity arose in 1939. By then, the Japanese army had suffered several defeats in battles with Soviet forces at Manchuria's borders. Ishii was now given permission to try his bioweapons against the Russians. A small force was sent to the border, where they spread many kilos of typhoid bacteria and other Salmonella bacteria on the banks of a border river. Additionally, shells prepared with various bacteria were fired at Russian forces. The results remain highly uncertain. Subsequently, several epidemics, including cholera, dysentery, and plague, broke out on both sides of the front line, resulting in numerous deaths among both Japanese and Russian soldiers. However, sanitary conditions were atrocious in the area, and the epidemics may have arisen naturally rather than as a result of bioweapons.

Nevertheless, Ishii managed to convince his superiors that the bioweapons had been a success in the battles with the Russians. He was given considerable freedom to continue testing such weapons in a series of attacks throughout Manchuria and China. He initiated epidemics of typhoid fever and cholera by contaminating wells. The plague bacterium was one of his favoured weapons, which could be spread in various ways. On several occasions, plague-infected rats were released in densely populated areas where epidemics broke out. In other instances, Ishii dropped packages of millet and wheat infested with plague-carrying fleas over cities. This also led to plague epidemics spreading in the vicinity as infected individuals fled the city.

Bioweapons were used less frequently during larger military operations. The most significant example of this is the Zhejiang campaign, which was triggered by the so-called Doolittle raid in April 1942 when American bombers led by Lieutenant Colonel James H. Doolittle attacked Japan for the first time and bombed Tokyo. The plan was for the bombers to land afterward in Zhejiang Province, which was the only coastal area in China not occupied by the Japanese. Most of the planes managed to do this. The Japanese now feared that this would be repeated through new American bombing raids against Japan. They attacked Zhejiang Province with a force of 175,000 soldiers to destroy airfields and other infrastructure.

During this campaign, the Japanese spread plague bacteria and other bacteria from the air. They further contaminated wells with cholera and other bacteria, and they scattered plague-infested fleas and anthrax bacteria in rice fields. Finally, they infected 3,000 Chinese war prisoners with typhoid and paratyphoid bacteria and released them to spread the infections. The Japanese bioweapon attack triggered epidemics, but the estimated death toll varies widely and is difficult to ascertain with certainty; it likely involves many thousands. However, the Japanese bioweapon attack also demonstrated that such weapons can be a boomerang. As many as 10,000 Japanese soldiers themselves became infected with the microbes used, and at least 1,700 died.[27]

The Zhejiang campaign is considered the most extensive example in history of the use of bioweapons on the battlefield. After this, the Japanese bioweapon attacks decreased. During the remainder of the war in China, only minor and half-hearted attacks were made. The background for this is unclear. However, there are some indications that the Japanese military leadership reacted to all the Japanese fatalities resulting from their own use of bioweapons. Perhaps there was also a lack of conviction regarding the military utility of this operation, which was very resource intensive. This might also explain why Ishii was reassigned to a smaller bioweapons centre in Nanking.

It is unknown how many deaths in China and Manchuria resulted from the Japanese bioweapon programme, with Ishii as a central actor. But the number is undoubtedly very large. It was primarily the civilian population that suffered. Some Chinese authors claim that 750,000 people perished, although this estimate may be too high.

It is remarkably difficult to assess the impact of the Japanese attacks using biological weapons, as the microbes used were also present naturally in China and Manchuria, causing epidemics. On the other hand, it is essential to emphasize that the use of such microbes can lead to the establishment of reservoirs of bacteria in local animal species, thus ensuring they remain in the area and can later cause infections and fatalities. This applies particularly to the plague bacterium, which, in many areas of Central Asia, exists as reservoirs in rodent species.[13]

As World War II drew to a close, Japanese leaders increasingly viewed defeat on the horizon. This led to several new desperate military measures, including thousands of kamikaze attacks – suicide attacks where pilots crashed planes loaded with bombs into Allied warships. A special type of kamikaze attack was planned by Ishii and some others, who still believed in bioweapons and aimed to use such weapons against the main enemy, the USA. This spectacular operation was given the poetic name 'Cherry Blossoms at Night' by Ishii.[28, 29]

The plan was for twelve submarines, each carrying three small planes with foldable wings, to cross the Pacific to the California coast near San Diego. The planes would then be launched using a catapult mechanism. They would fly over San Diego, San Francisco, and Los Angeles, dropping bombs laden with plague-infested fleas, possibly along with other microbes, over the cities. The kamikaze pilots would then crash their planes. This attack was never carried out. The supreme commander of the Japanese General Staff prohibited the operation, which was to occur in September 1945. On 2 September, Japan signed the capitulation agreement.

It is uncertain whether this particular kamikaze attack would have succeeded. Some deaths from the plague would likely have occurred, but the extent of any potential epidemic would have been far less than the Japanese plague attacks in China due to differences in medical resources and epidemic control. However, such an attack would have created immense fear among the American population and simultaneously represented a significant propaganda victory for Japan.

In the final months before capitulation, there were also plans to send balloons carrying bioweapons over the Pacific toward North America. This project also did not materialise. Over a thousand balloons filled with explosives were sent and reached the USA but caused little damage. Attacks using bioweapons could have been carried out in this manner as well.

When Japan's defeat became a reality and Soviet troops advanced into Manchuria, the Japanese blew up the laboratories at Unit 731. Remaining prisoners and witnesses were liquidated, and prisoner records were destroyed. A similar 'clean-up' at Unit 100 in the capital Changchun was concluded in a characteristic way: 60 horses infected with glanders were released, along with hordes of rats infected with plague bacteria and other microbes. This triggered epidemics that made the surrounding area uninhabitable for the next decade.

Researchers from the Japanese bioweapon centres returned to Japan. Ishii advised them to keep a low profile in the time ahead.

The Americans' Pact with the Devil

Even during World War II, the Allies received reports about the Japanese's ruthless use of bioweapons in Manchuria and China. They were often dismissed as unreliable Chinese propaganda and given little weight. Nevertheless, by the end of the war, there was a clear perception in the American intelligence community that Japan had an extensive bioweapon programme from which valuable insights could be gleaned. The USA also wanted to establish such a programme.

By the end of the war, the relationship between the Soviet Union and the Western Allies had begun to deteriorate. There was increasing fear that the Russians had a bioweapon programme. Soviet forces had occupied parts of Manchuria, including Unit 731. Were the Russians now acquiring useful information from Japanese bioweapon research? It was crucial to get ahead of them.

Shortly after Japan's capitulation, researchers from Camp Detrick, which was the US centre for biological weapon research, came to Japan to interview Japanese scientists linked to the bioweapon programme. The Japanese, who had destroyed much written material, asserted that their programme had been exclusively defensive, aimed at developing defences against biological weapons.[30]

Initially, little came of these interviews. However, the Americans continued to harbour a nagging suspicion that they could gain more valuable insights into the Japanese's experiences with bioweapons and continued the investigation. In early 1947, they finally interviewed Ishii himself, who gradually became cooperative. Crucially, Americans secretly promised both Ishii and several other key figures from the bioweapon work amnesty – full immunity from prosecution. It is reasonable to believe that President Harry Truman supported this.

This was the main reason the Americans agreed to such an arrangement: General Douglas MacArthur, now the occupation chief in Japan, believed that the public disclosure of the barbaric actions at the Japanese bioweapon centres could create significant unrest in a country that was now being rebuilt as a democracy – and, at worst, lead to attacks on American occupation troops. To prevent this, the emperor had to be shielded from any suspicion of responsibility for Japan's wartime actions. Since the emperor was very likely aware of the bioweapon programme, war criminal trials against Ishii and his associates would also implicate the emperor.[30]

The result of these considerations was extensive, collegial discussions between American and Japanese scientists. Ishii and his colleagues could openly and without fear recount everything about their operations, including the horrific experiments with human test subjects. From the extensive reports produced, it seems that the American scientists viewed the Japanese results they obtained as ordinary scientific observations. There are no signs of ethical reflections on the Japanese's barbaric research methods. British intelligence was also kept informed about the results of the interviews with Ishii and his colleagues.

The decision to conceal all information regarding Japan's bioweapon engagement and the negotiations with Ishii and his people stemmed not only from conditions in Japan. There was also an awareness that revealing the truth about the Japanese's barbaric activities would provoke an intensely strong backlash in the West, especially if the perpetrators went unpunished. This could lead to such strong opposition to bioweapons that it would hamper the development of the US programme. There was indeed a certain openness regarding these plans in the first few years following World War II.[30]

The outcome was that the world did not learn anything about Japan's bioweapon programme and the shocking details surrounding it. The Japanese scientists responsible were left completely free. Some became professors of medicine, while others held high positions in research or industry. Little is known about Ishii's activities after 1948. Some have claimed that he was in the USA for a period, while others say he resurfaced during the Korean War. He died of cancer in 1959.[28]

Initially, Americans concluded that the Japanese had not achieved significant results concerning the technology for spreading bioweapons effectively. However, they believed that the autopsy results from human trials might be of interest, as they, of course, did not have access to conduct similar experiments on humans in the West. But upon closer examination, it turned out that these results were also of little value. The experiments Ishii and his colleagues conducted were partly unsystematic and featured experimental conditions that were unrealistic for the practical use of biological weapons.[30]

The concealment of Japan's bioweapon programme was highly efficient – both in the USA, the UK, and Japan. Neither during the International Military Tribunal that convened in Tokyo in 1946 – termed Japan's 'Nuremberg Trials' – nor during a number of other trials against Japanese war criminals was this inhumane programme highlighted.

In the Soviet Union, however, twelve Japanese individuals associated with the bioweapons centres in Manchuria and China were put on trial in 1949 in the Siberian city of Khabarovsk. The defendants gave full confessions regarding their activities at Unit 731 and Unit 100 –including the experiments on humans, of which both the USA and the UK had been informed. However, the authorities in the USA and the UK dismissed all revelations from the Khabarovsk trial as Stalinist propaganda. The Japanese defendants received surprisingly short prison sentences, suggesting that they also made some sort of deal with the Russians similar to that of the Americans.[30]

It would take several decades before the world finally learned about the horrors of the Japanese bioweapon programme. Some of the individuals responsible eventually came forward and confessed to their crimes. One of them sought to come to the USA to atone for his crimes by giving lectures on the bioweapon programme at Unit 731. As a final ironic reflection of America's double standards, he was denied entry – after all, he had confessed to war crimes![28]

It has been suggested that the silence surrounding the Japanese bioweapon programme may have contributed to many countries subsequently keeping their own development of such weapons secret in the following years. Public opinion in the post-war period was primarily concerned with the threat posed by nuclear weapons and much less with bioweapons.

The Red Army Invests in Bioweapons

During World War I, the Russian army was severely affected by the Germans' surprising use of chemical agents, which caused thousands of soldiers to lose their lives. The Bolshevik regime that came to power with the revolution in 1917 was determined to build a modern army equipped with all types of weapons. They therefore focused on chemical agents. However, the Bolsheviks also sought to develop biological weapons.[31, 32]

Both during World War I and in the subsequent civil war between the Reds and Whites, more people in Russia died from various infectious diseases than from combat injuries.[13] In particular, typhus was seen as a problem. From 1918 to 1921, it is estimated that 25 million in the civilian population and the Red Army were infected with typhus, leading to 3 million deaths. A third of the doctors in the Red Army died. The problems became so severe that Vladimir Lenin reportedly stated: 'Either socialism will defeat the louse, or the louse will defeat socialism.' The typhus bacterium is, after all, transmitted by lice. The devastating impact infectious diseases had on military forces likely contributed to the Bolsheviks' desire to develop such weapons.

It was Jakov Fishman who became the head of the chemical weapons development programme in 1925. He was also very interested in biological weapons and is considered the founder of the Soviet bioweapons programme. He established a laboratory in Moscow, which began studies of anthrax and botulinum toxin as weapons. Cholera and plague bacteria eventually became part of the repertoire as well. Fishman's activity was tied closely to the Red Army.

It is also claimed that the OGPU – the powerful security police of the Soviet Union and a precursor to the NKVD – conducted its own studies of bioweapons, including the typhus bacterium, independent of Fishman's activities.

We do not have secure information about what actually transpired in the field of bioweapons in the Soviet Union during the early years. However, we know that Fishman, in 1928, prepared a detailed report for the authorities on bioweapons, presenting the state of the field and making recommendations for future efforts. This led to the establishment of an extensive organisation for a Soviet bioweapons programme in the same year. Several research institutions at universities, military institutions, and OGPU laboratories were part of this programme, which was under Fishman's leadership but, of course, under government oversight. Centres for field research into biological weapons were also established. The most significant of these was established on Vozrozhdeniya Island in the Aral Sea, in what is now Kazakhstan. Here, extensive field studies of various bioweapons were eventually conducted. As we will see later, this had serious consequences for the local population.

There is still much uncertainty surrounding the Soviet bioweapons programme during the interwar period. Nonetheless, it is reasonable to conclude that interesting results were achieved for potential military applications of microbes. General, later Marshal, Mikhail Tukhachevsky – a gifted military theorist and strategist – was primarily responsible for modernising the Red Army. He was also favourable to unconventional weapons, which may have benefited the bioweapons programme. Many capable researchers were engaged in this effort. In addition to developing microbe weapons, there was also significant focus on defences against such weapons – there was considerable fear that Germany and several other nations were active in this field. Russian researchers strongly emphasized the development of vaccines against numerous bacteria, such as anthrax, plague, tularaemia (hare plague), brucellosis, and tuberculosis.

However, the Soviet bioweapons programme suffered a severe blow when the Great Purge of 1937 affected politicians, military personnel, and scientists.[31] This was an extensive cleansing process, carried out by the NKVD, initiated by Joseph Stalin to eliminate anyone he perceived as a threat based on his paranoid delusions. The trials were based on falsified evidence and torture. A long list of prominent microbiologists and heads of research institutions were accused of treason, including planning sabotage operations using bioweapons in collaboration with foreign powers like Japan and Germany. Several of the scientists were executed, while others were imprisoned or exiled to Siberia. The father of the bioweapon programme, Jakov Fishman, was imprisoned in 1937 and only released in 1954. Marshal Tukhachevsky, who may have been a high-ranking protector of the bioweapons programme, along with several other top leaders of the Red Army, was put on trial based on absurd charges and shortly afterward executed by firing squad.

Some of the imprisoned microbiologists who were of particular interest to the NKVD were placed in special work groups – called Sharagy – where they could continue work relevant to the bioweapons programme.

Typically, it was the most skilled scientists in biological weapons who fell victim to the Great Purge, while many others went free. This undoubtedly weakened the Soviet bioweapons programme. Nevertheless, there were likely enough researchers to continue work on defensive measures and probably also on the development of offensive weapons. After the Soviet Union's entry into World War II, this weapon activity increased. There is some evidence that human experimentation was conducted in prisoner camps and prisons. The subjects were reportedly infected with plague, anthrax, and glanders bacteria. However, this is not documented in Russian sources.

In a speech in 1938, Defence Minister Marshal Kliment Voroshilov stated that the Soviet Union possessed biological weapons and was prepared to use them in retaliation if attacked with such weapons. It is unclear what was based on reality and what was propaganda in this speech.

In 1939, Stalin appointed his notorious interior minister, Lavrenty Beria, as the supreme chief of the bioweapons programme. This suggests a genuine commitment to this field. Nevertheless, it is believed that the Soviet Union did not use bioweapons during World War II. Some have claimed that biological weapons were deployed at least once – during the historically significant Battle of Stalingrad.

As the Germans' 6th Army under General Friedrich Paulus advanced towards the city in July 1942, a Russian defector, Ken Alibek, reported that thousands of German soldiers became ill with tularaemia, allegedly slowing their advance.[33] Shortly thereafter, a significant epidemic developed, also affecting the civilian population in the area.

The majority of the German soldiers with tularaemia, according to Alibek, had pneumonia. This is a rare form of tularaemia, which may indicate that the bacteria had been aerosolised by Soviet forces, who were in a desperate situation.

Many experts, however, doubt this claim.[7] Earlier that year, there had been a significant naturally occurring tularaemia epidemic in the area. This new outbreak could have ecological causes related to the war situation. German military reports also suggest that their soldiers were infected through natural means, that fewer soldiers were affected, and that the Germans' advance was not impeded significantly. We have no definite explanation for this tularaemia outbreak. Regardless, it was likely the Red Army with conventional weapons – and not the tularaemia bacteria – that caused the Germans' fatal defeat at Stalingrad, one of the major turning points of World War II.

French Fear of the Arch-Enemy

The Treaty of Versailles imposed strict limitations on military rearmament in Germany. Nevertheless, there was significant fear in France that the Germans would attempt to circumvent these provisions.[25,34] This fear was hardly unfounded, as the German desire for revenge was strong.

As part of the monitoring of Germany, the French scientist Auguste Trillat was sent in 1919 to a large chemical laboratory in Germany. Trillat was the head of a chemical research institute connected to the navy. One of the directors at the German laboratory he visited confided to Trillat that efforts to develop bioweapons were continuing in secret. It had become known that Germany had used bioweapons significantly during the war, including against France. Trillat prepared a comprehensive report on his findings and opinions regarding bioweapons, clearly stating that France should invest in such weapons. Subsequently, the French government decided to initiate a bioweapon programme, although officially the focus was primarily on defence against biological weapons.

The intensity of the French bioweapon programme fluctuated in parallel with the tense relationship with Germany and fears of war. The first active phase was from 1921 to 1926. Auguste Trillat became a central figure in France's efforts in biological weapons.[34] The prestigious Pasteur Institute in Paris, where Trillat had played a leading role for several years, became the centre of French bioweapon research, although several other major laboratories were also involved. Many different aspects of bioweapons were studied: selection of microbes and toxins, methods of dispersal, and various means of defence against bioweapons.

Trillat had already conducted important studies on the transmission mechanisms of respiratory infections. He was particularly interested in the significance of what he called 'clouds of microbes', wherein the clouds consist of such small droplets that they remain suspended for a long time – what we today refer to as aerosols. In this field, he was a pioneer. As we have seen, the dispersal of microbes as aerosols is crucial in the context of bioweapons. Trillat determined that such 'clouds' could be most effectively produced with microbe-carrying bombs or grenades, whereby the explosion results in the formation of clouds of very small particles that remain in the air for a long time. In the following years, experiments were conducted to perfect such bombs. Various bacteria and botulin toxin were tested as bioweapons, with both animal experiments and trials involving aerial bombs carried out.[34]

Trillat also had the original idea that 'clouds' – aerosols – composed of disinfectants, such as sulfuric acid or chlorine, could provide protection against bioweapons, and conducted experiments to study their effects. Different forms of vaccination against relevant microbes were also examined, including animal experiments with mice vaccinated with microbe-containing aerosols.

From 1927 to 1934, the French bioweapon programme entered a more passive phase. This was likely due to France's desire to adhere to the provisions of the Geneva Convention from 1925. The fear of Germany had probably also waned somewhat. However, Trillat and several others continued their work on bioweapons regardless.

In 1933, the Nazis, led by Adolf Hitler, came to power in Germany. The same year, Germany withdrew from international disarmament efforts and immediately exited the League of Nations. War fears increased in France, leading to a decision to resume bioweapon work with greater intensity than before. Both offensive use and defence against bioweapons were emphasized.[25, 34]

Work on microbe aerosols continued as well. Experiments were conducted where aerosols containing a supposedly harmless bacterium – *Serratia marcescens* – were tested in the Paris metro to study the extent of dispersal. We now know that this bacterium can cause serious infections, especially in patients with compromised immune systems. There are no reports of such infections following the testing in the metro, but no follow-up investigations were conducted either.

Additionally, animal experiments were carried out to study whether projectiles containing anthrax spores could cause wound infections. The activity in the French bioweapon project further intensified in the months following the outbreak of World War II. In the last months before Germany invaded France, the French worked tirelessly on developing effective bombs and grenades with bioweapons. Field tests were conducted until April, just weeks before France's collapse. It is reasonable to assume that these trials yielded very useful results, but the French chose to destroy all documents to prevent them from falling into German hands.

It is likely that the French continued bioweapon trials until 1942 in the part of France that was not German occupied. However, little is known about this. Regardless, it is clear that the long-standing French bioweapon programme, which could undoubtedly point to a number of interesting trial results, never led to France using biological weapons during the war.

Germany and Bioweapons – Paradoxes and Enigmas

With its active use of bioweapons during World War I, Germany is, apart from Japan, the only state that can be said to have engaged in biological warfare in modern times. Given this background, and considering the massive microbiological expertise that existed in Germany during the interwar period, one might expect that the country would have developed an extensive bioweapon programme as part of its military rearmament. Rumours that this was indeed occurring were one of the reasons that France, among others, invested in bioweapons. After the outbreak of World War II, there was significant fear in Britain and the USA that Germany had both bioweapons and the capability to use them.

The realities were entirely different, but this was only made clear after the war ended. When Germany entered the war, it had no proper bioweapon programme and was not capable of using bioweapons in its warfare. The background for this is complicated and only partially clarified. Several factors were likely decisive.[24]

In the years following World War I, interest in bioweapons was minimal in Germany. There were rumours that both France and the Soviet Union were investing in this type of weapon, which led to some meetings between experts and politicians. However, this did not result in systematic work on developing such weapons. Military experts in the field believed that biological weapons had little practical utility and could even be dangerous for the side employing them. The interest in chemical weapons was far greater, as there was ample experience from World War I.

When Germany launched a blitzkrieg invasion of France in the spring of 1940, it gained access to a wealth of documents from one of the centres of the French bioweapon programme, which showed how far the work had progressed. A leading German expert reviewed the material and prepared a report concluding that Germany should also invest in bioweapons. This did not initially lead to anything. However, reliable information about the Allies' active work on such weapons gradually became available. Several military leaders were now convinced that Germany should develop bioweapons, including the head of the army high command, Field Marshal Wilhelm Keitel, who had been one of the sceptics previously.

Adolf Hitler – The Main Obstacle

However, one insurmountable barrier prevented Germany from launching a full-scale active bioweapon programme that could match the Allies' activity in the field. Adolf Hitler – the undisputed Führer of Germany – staunchly opposed the use of bioweapons in wartime.[24] Instead, he ordered that all efforts should be focused on measures to defend against such weapons. In this matter, Hitler was unyielding. Frustrated military leaders and researchers repeatedly tried to convince him that the time had come to develop bioweapons, but to no avail.

What was the background to Hitler's intense aversion to the use of biological weapons? There is no convincing explanation for this, and Hitler himself never provided a reason. A number of more or less plausible theories have been suggested. It has been claimed that Hitler opposed bioweapons because he was injured by mustard gas during World War I. This is unlikely. After all, he was a strong supporter of investing in chemical weapons and considered the nerve gas tabun his 'miracle weapon', which could potentially be dispersed using bombs.

It is more likely that he feared retaliatory attacks with bioweapons if Germany were to employ its own. On the other hand, a German bioweapon arsenal could also have a deterrent effect, as was seen with the use of chemical weapons during World War II, where both sides had such weapons but refrained from using them. Nonetheless, Hitler firmly rejected the development and use of bioweapons.

More irrational reasons may have influenced Hitler's views on biological weapons. According to his physician, he had a pronounced phobia against bacteria. This may have been linked to his fanatical anti-Semitism, where 'the Jew' was regarded as impure and was even compared to dangerous pathogens for the societal organism.

Despite Hitler's stance, attempts were made in 1943 to organise and coordinate work on bioweapons.[24] Keitel established, likely with Hitler's consent, a committee consisting of experts from the military divisions primarily tasked with defensive measures against biological weapons. It was code-named Blitzableiter – 'Lightning Conductor'.

However, Hitler's military advisors were dissatisfied with the restriction of this work to bio-defence. The powerful leader of the SS, Heinrich Himmler, and the commanding officer of the Luftwaffe, Hermann Göring, chose to establish another secret bioweapon group that would focus on the offensive use of bioweapons if the Führer changed his mind during the course of the war. Göring appointed a professor of medicine, Kurt Blome, as the head of this group. Blome had no experience in the field and had worked previously on cancer research, which would now be a cover for the group's activities. Keitel was aware of this, but evidently did not want to be associated directly with Blome's activities, which he expected would involve dubious experiments on humans.

Himmler was extremely interested in bioweapons. He promised Blome full support and access to SS resources, including the use of prisoners in concentration camps for experiments. The centre for Blome's bioweapon work was the so-called Central Institute for Cancer Research (Zentralinstitut für Krebsforschung), established in Posen (Poznan in present-day Poland). The plans for this facility were ambitious, including various buildings and medical and scientific facilities. Despite Himmler's promises, and to Blome's great frustration, construction progressed slowly. Blome later claimed that no work was done there. Nevertheless, there are reasons to believe that experiments on humans were conducted, including trials of plague vaccines on concentration camp prisoners. As the Red Army approached Posen at the end of 1944, all operations were relocated to the west, and a new facility was started in Geraberg in Thuringia. Construction had only just begun when the American army moved into the area in April 1945.

Thus, two independent organisations had been established to intensify Germany's delayed bioweapon programme – the Lightning Conductor organisation and what has been called the 'Blome network'. This approach to tackling the problem was representative of the organisational structure in the Third Reich. Political control was divided among a number of different bodies that operated independently of one another – and competed extensively with each other. Hitler clearly desired such a structure. The prerequisite was, of course, that the work was done in accordance with his ideology – with the 'will of the Führer'.[24]

There was no coordination of efforts between the two bioweapon organisations, and there was virtually no collaboration. Both organisations included a number of different centres, but these were small and had limited personnel and resources. Many of the scientists responsible for the projects were likely not especially skilled. Blome had, as mentioned, no relevant academic background in bioweapon research, and one of those who interrogated him after the war described him as 'a complete fool'. However, there were also some exceptions. One of them, Eugen von Haagen, was a highly skilled virologist who in 1936 was nominated for the Nobel Prize for his contribution to the development of a vaccine against yellow fever. He was now working at a centre in Strasbourg on developing a vaccine against typhus and conducted numerous trials on concentration camp prisoners whom he had infected with typhus bacteria.

These conditions meant that there were no significant breakthroughs in Germany's bioweapon efforts during the war that could have enabled the Third Reich to employ them. However, it is clear that both of the aforementioned organisations – Lightning Conductor and the Blome network – were working on aspects related to the offensive use of bioweapons, not just defence against them. Their circumvention of Hitler's clear and repeated orders to focus solely on defensive measures was justified by stating that this work in practice required experience with bioweapons. This is a dilemma that always arises in bioweapon research and can very well be utilised as a pretext for working on offensive bioweapons.

As mentioned previously, Japan had a comprehensive bioweapon programme and used these weapons extensively in its warfare in China and Manchuria. Japan was allied with Germany. Did they collaborate on bioweapon development? There is little evidence to suggest that they did.[24] On a few occasions, contacts were made regarding cooperation, but these did not lead to concrete results, although some have claimed otherwise. This lack of mutual information between Japan and Germany was evident in other areas as well. Germany did not inform about the attack on the Soviet Union in 1941, and Japan did not warn about the attack on Pearl Harbor the same year. Germany was apparently also not informed about Japan's use of bioweapons in China.

Ruthless Experiments

What we know about the work conducted on bioweapons in Germany during the war years is that it was partly carried out with the same ruthlessness and disregard for human life that characterised the medical experiments in concentration camps. Prisoners from concentration camps were used, among other things, for the development of various vaccines. Although the practical results largely failed to materialise, there were extensive plans for bioweapon use. Both humans and animals were considered as targets.

For example, experiments were conducted with the virus that causes foot-and-mouth disease in cattle. This was intended for use against England to incapacitate key parts of the agricultural sector.

There was also work on aerosols with various microbes, and some experiments were conducted with microbe bombs from aircraft. Other examples include the work at the research institute for entomology (insect research) linked to the concentration camp in Dachau.[36] This was established by SS chief Himmler and was part of Kurt Blome's 'empire'. The work there aimed partly at defence against insect-borne infections, but they also investigated the potential use of insects as weapons. Among other things, experiments were conducted with malaria-infected mosquito species on Dachau prisoners. The spread of plague using rats was also an area of interest. Blome allegedly had the idea that plague-infected rats could swim ashore from submarines on the English coast, but experiments showed that the rats drowned quickly without seeking land.

Overall, much of Nazi Germany's work on biological weapons was characterised by aimlessness, lack of coordination, and a degree of amateurism. There was also a scarcity of resources at the many smaller centres responsible for the undertaking. All this contributed to the lack of practical results. But ultimately, it was Hitler's negative attitude toward the offensive use of bioweapons that was decisive, even though attempts were made to circumvent his prohibition.[24]

It is not unlikely that the German bioweapon programme – despite its organisational weaknesses and late start – would have had greater success if it had the Führer's wholehearted support. If Hitler had had effective bioweapons, would he have used them in the final phase when Germany was backed against the wall? Would bioweapons have been part of Hitler's visions of *Götterdämmerung* while he sat in his bunker beneath the Reich Chancellery? This remains pure speculation.

Cynical Aftermath

The history of the failed German bioweapon programme has a sequel reminiscent of what happened to Ishii Shiro after Japan's capitulation. The Americans were highly eager to enlist German scientists in their service, partly to gain insights into any valuable research results –and not least to prevent the Russians from obtaining them. This was part of a project called Operation Paperclip, whereby a total of 1,500 German and Austrian scientists and their families were granted asylum in the USA.[37]

In their eagerness, the Americans chose to overlook uncomfortable facts: several of the scientists in question had not only been members of the Nazi Party but had likely also committed war crimes and should have been brought to trial. This applied to a number of scientists from the bioweapon programme, including Kurt Blome. He was indicted during the so-called Doctors' Trial that the Americans conducted in Nuremberg in 1946 and 1947, where German doctors were put on trial for war crimes. Surprisingly,

Blome was acquitted, likely under American pressure. This probably saved him from the gallows. American intelligence believed he could provide very valuable information about both bioweapons and chemical weapons, which he had also worked with. He was hired as an expert by the Americans for several years but was not granted a visa to enter the USA. Later, he worked for many years as a doctor in West Germany and was politically active. He died in 1969 without ever being brought to justice after his undeserved acquittal in Nuremberg.[35]

Churchill and the Anglo-Saxon Brothers-in-arms

When World War II broke out in September 1939, only France among the Allies had a bioweapon programme.

In 1934, British journalist Henry Wickham Steed attracted attention with alarming claims. He alleged that he had gained access to secret German plans showing that Germany had made significant progress in developing airborne bioweapons. He also claimed that the Germans had conducted experiments with aerosol dispersion of bacteria in the metro systems of both Paris and London. Officially, both the British and French governments expressed doubt about the validity of Steed's claims, but behind the scenes, they were taken seriously. They coincided with an intelligence report that had recently been received, also indicating strong German involvement in the development of biological weapons.

The eventual decision by Britain to invest in a bioweapon programme was likely due to one man – Colonel Sir Maurice Hankey, later Lord Hankey. He has been referred to as the father of the British bioweapon programme.[34, 38] Hankey had, as a shadowy figure behind the scenes, influenced British defence policy significantly since World War I. As a government secretary, he had an extensive network that he now mobilised to examine the necessity of building up a British bioweapon programme. Although he had considerable aversion to such weapons, which he referred to as 'these forms of frightfulness', he contacted a group of prominent scientists, particularly bacteriologists, and established a committee for bioweapons with himself as chairman. The committee produced several reports that initially expressed some doubt about the significance and utility of bioweapons. Nonetheless, the committee recommended strengthening defences against potential attacks with such weapons.

The Nobel Prize Winner Who Developed Bioweapons

In November 1939, London welcomed a Canadian visitor who heightened British interest in bioweapons. It was the world-renowned scientist Sir Frederick Banting, who came to present his views on biological weapons and the necessity of investing in them to the British authorities. Banting had won the Nobel Prize in 1923, along with John Macleod, for the discovery of insulin.[39] He was a war hero from World War I and passionately interested in bioweapons, which he believed were absolutely necessary for the Allies to be able to respond to similar attacks from Germany and the two other Axis powers – Japan and Italy. As early as 1937, he had begun working on plans for bioweapon development on his initiative, despite lacking any academic background in microbiology. He established contact with a number of outstanding scientists at several Canadian universities and presented his plans to the Canadian authorities.

Upon arriving in England in November 1939, Banting had meetings with several key figures, including Hankey, where he presented his views on bioweapons – relevant microbes, practical weapon development, and testing. He suggested that bioweapon dispersal from aircraft, which he was very concerned about, could be conducted in Canada, which had vast, uninhabited areas suitable for this.

Great Britain goes for Biological Weapons

Banting left England feeling that he had not convinced the British and was fairly bitter. He described Hankey as 'an excellent example of the servile, self-important, self-satisfied, supercilious donkeys in the British government'. Hankey held a ministerial rank in the government at that time. However, Banting was mistaken as Hankey acted on his proposal. Despite facing some resistance – even within his own committee – it ultimately resulted in the government under Neville Chamberlain deciding to actively invest in bioweapon research, also considering the offensive use of such weapons. It was decided that a dedicated centre for this work would be established at Porton Down in Wiltshire, where there was already a centre working on chemical weapons.[34, 38]

Hankey ensured that the eminent bacteriologist Paul Fildes became the head of the centre at Porton Down. Fildes was a hard-working bachelor with a particular sense of humour. He was arrogant, determined, and somewhat authoritarian. He quickly chose to concentrate on two types of bioweapons that he believed were most promising: anthrax spores and botulin toxin.

Winston Churchill, who became Prime Minister in May 1940, also supported the intensified work on bioweapons and was kept informed about the work at Porton Down.

Given the urgency of developing a biological weapon that could be used in response to a potential German biological attack, Fildes and his team initially decided to target the cattle population in Germany with anthrax. Since both animals and humans can develop anthrax after ingesting spores through the mouth, he developed so-called anthrax cakes – seed cakes in which deadly amounts of anthrax spores were baked. The cakes themselves were produced in large quantities by a soap factory in London. At Porton Down, the cakes, which had a hollow centre, were filled with anthrax spores using specialised machines and sealed so that they posed no risk during handling. This operation was code-named Operation Vegetarian.

In the area around Porton Down, experiments were conducted with cattle, sheep, and horses to assess how the animals reacted to the cakes. Most of the trials were carried out with 'unloaded' cakes, but a few involved anthrax. The cakes worked as intended. A stockpile of 5 million anthrax cakes was built up. Fildes had calculated that dropping this bomb load over the cattle areas in north Germany would decimate large parts of the cattle population and would impact food supplies in Germany. It was assumed that only animals would be affected. However, the reality is that such an anthrax attack would also likely result in illness and death among humans. However, the anthrax cakes were never used and were destroyed after the war.

Interestingly, a British UN delegate in 1969 stated that 'as several subsequent governments have stated, we have never had any biological weapons'. This clearly inaccurate comment illustrates that, when it comes to bioweapons, one can never expect truthful information from the authorities.

Fildes saw his primary task as developing an anthrax bomb for aircraft as an anti-personnel weapon. To achieve wide dispersal, the anthrax spores needed to be in aerosol form. Fildes and his team conducted careful experiments in the laboratories at Porton Down with small test animals in specially designed chambers to study the effects of anthrax aerosols and find suitable doses of spores. Next, it was necessary to conduct trials with anthrax in open conditions that were more realistic for dispersal via aircraft bombs.

Gruinard Island – The Anthrax Island off Scotland

Realistic experiments with anthrax bombs on larger animals could not be conducted at Porton Down due to safety concerns, as nearby village populations could be affected. The experiments required a more remote location, and a suitable area was found on the small uninhabited Gruinard Island just off the north-west coast of Scotland. A team from Porton Down travelled there in the summer of 1942.[38]

Approximately 150 sheep were purchased from the mainland population. The experiments with anthrax bombs involved tethering a number of the animals a certain distance apart over an area of several hundred metres. Bombs, which were suspended in a kind of gallows 1.2 metres above the ground, were detonated remotely, and the effects on the sheep were recorded. After the explosion, a cloud of anthrax aerosol was formed. It turned out that the vast majority of sheep within at least 300–400 metres of the explosion died after a few days.

To conduct a trial with anthrax bombs from aircraft, a remote beach in Wales was chosen, where bombs were dropped from an altitude of 1,500 metres. All sheep within a distance of up to 400 metres from the bomb impact died.

Further trials were conducted both on Gruinard Island and at Porton Down to study the necessary doses of anthrax spores in the bombs and various bomb variants. At Porton Down, the experiments were done with spores from a harmless bacterium, *Bacillus subtilis*.

Fildes concluded that all these tests showed that an anthrax bomb could indeed be a useful military weapon. At the same time, he was convinced that Germany had also developed a similar weapon. Therefore, it was crucial to mass produce such bombs for potential retaliatory strikes. However, in

Britain, this created problems. Several key individuals were sceptical of Fildes and his anthrax bombs. They doubted not only the military utility of bioweapons but also believed there were insufficient resources for the development and production of such weapons, which would detract from more critical wartime objectives.

The Triangle Cooperation Takes Off

The way out for Fildes and his supporters, including Lord Hankey, was collaboration with the USA, which was very interested in the results from Porton Down that Fildes had been informing regularly about. Ultimately, a close triangular cooperation on bioweapons developed between Britain, the USA, and Canada.[34, 38]

Frederick Banting had been the driving force behind establishing a bioweapon programme in Canada, in fact before both Britain and the USA. He had organised a network of skilled scientists both at his own research laboratory at the University of Toronto and at several other universities, and he received support from central figures in the industry. He had also initiated cooperation with the USA.

As we have seen, Banting also attempted to motivate the British authorities to invest in bioweapons during his visit to England in late 1939. After returning home, he initially received only limited support from the authorities for his work on such weapons, but he continued unfazed. In 1940, he began studying the possibilities of dispersing microbes from aircraft, initially using sawdust that he later intended to mix with bacteria. He was very enthusiastic about the experiment, even though it has been somewhat maliciously claimed that it 'showed little more than that gravity works'. He had many ideas for further work, but Canada lacked both the resources and personnel necessary to proceed alone. However, Banting had established valuable contacts in the USA who were interested in his views and potential collaboration.

In February 1941, Banting died in a plane crash while en route to England to discuss further cooperation on bioweapon development. This set back the Canadian efforts in the field somewhat, but the work was resumed after a time with the establishment of a closer collaboration between Britain, the USA, and Canada.

The most resourceful of the Anglo-Saxon triangle cooperation – the USA – was also the one that got started the latest.[34] When World War II broke out, the USA had no bioweapon programme as during the interwar period the military viewed such weapons as irrelevant. Nonetheless, by the late 1930s and within the early war years, fears spread that both Japan and Germany might resort to using this type of weapon, and there were rumours that this was already happening in China. In 1940, a meeting was held regarding bioweapons in Washington with participation from British and Canadian representatives, although nothing resulted from it initially.

Not until late 1941 was a separate committee established in the USA by central scientists and military officials to address the bioweapon issue. They produced a report that concluded biological weapons were an imminent threat that needed preparation against and that active studies and possibly production of such weapons should commence. President Franklin D Roosevelt endorsed this, and a bioweapon programme was established with broad participation from research communities at several universities and pharmaceutical companies. The programme was strictly secret and was led by George Merck, the head of one of the largest pharmaceutical companies in the USA.

The centre for bioweapons research was established at Camp Detrick, later renamed Fort Detrick, in Maryland, where work commenced in the spring of 1943. Several other centres were also established in collaboration with Camp Detrick.

In the autumn of 1942, Paul Fildes and a colleague visited both Canada and the USA, presenting their findings on anthrax bombs. This undoubtedly made an impression on their American and Canadian colleagues. This led, among other things, to the further development and refinement of anthrax bombs becoming the central project at the newly established centre at Camp Detrick. There, thorough studies involving animals were conducted to ascertain the conditions for the most effective infection with aerosols in dedicated chambers. However, primarily, they continued to test and improve various bombs through extensive field studies with aerial weapons at their facilities. Gradually, plans were made for mass production of suitable bomb models. There was significant pressure for this from British authorities.

Meanwhile, Fildes and his team continued anthrax testing on Gruinard Island until August 1943, when they had to cease their work. The sheep that had perished during the anthrax trials had been buried near the beach on the island. Now, a violent storm had resulted in some of the carcasses being unearthed and washed ashore to the mainland, leading to fifty anthrax deaths in the sheep population. To calm the panicked local population, the government launched an explanation that the anthrax outbreak resulted from a Greek ship disposing of some infected sheep carcasses overboard. Fildes realised that the safest course would be to conduct further testing of anthrax bombs in Canada and the USA.[34]

In December 1943, Fildes could inform his superiors that the USA could commence mass production of anthrax bombs. This came as an unwelcome surprise to some key figures in Britain who, for various reasons, opposed the bomb project. The matter ended up on Churchill's desk. After consulting with his closest advisors, he greenlit the project. On 8 March, he decided that Britain would order 500,000 anthrax bombs from the USA as an initial measure. The bombs would eventually be transported to Britain for possible future use.

By the summer of 1944, Fildes worked on plans for using anthrax bombs against Germany. The plan was to simultaneously drop 4 million bombs over six German cities: Berlin, Hamburg, Stuttgart, Frankfurt, Wilhelmshaven, and Aachen. He estimated that this would result in 3 million anthrax deaths. The areas would be uninhabitable for years due to anthrax spores in the ground that could continuously cause new infections.[34]

However, the production of bombs in the USA was delayed significantly, and there were no deliveries to Britain as planned. The anthrax bombs were never used during World War II.

The trials conducted by Fildes and his team on Gruinard Island had a follow-up that highlights a significant issue with using anthrax as a bioweapon: the spores can persist in nature for years. When investigations revealed that the soil on Gruinard Island was full of anthrax spores, the island was purchased by the British government and declared a prohibited area for animals and humans. When investigations in 1986 still revealed an abundance of spores, thorough disinfection with formaldehyde was initiated over several weeks. In 1987, 40 sheep were allowed to graze on the island.

When none of the sheep contracted anthrax, Gruinard Island was declared free of infection. In 1990, the original owner was allowed to buy the island back, 47 years after Fildes' anthrax bomb trials.[40]

In the triangle collaboration between Britain, the USA, and Canada, anthrax was the only bioweapon that was considered. As mentioned previously, Fildes was initially also concerned with botulin toxin but quickly lost interest in this as a bioweapon. The USA and Canada, however, continued to work on botulin toxin, and the Americans believed such bombs could be promising. At Camp Detrick, work was also conducted on a range of other microbes that later became relevant as biological weapons, but there were no practical results from this work by the end of the war.[34]

Attitudes Toward Bioweapons During World War II

While Japan used bioweapons recklessly during its warfare in China and Manchuria, such weapons were not employed by the warring powers in the West. The reasons for this are many.

Since 1923 there had been an international agreement – the Geneva Protocol – that prohibited the 'first use' of bioweapons but apparently allowed the application of such weapons in retaliation. Unfortunately, there is little to indicate that this would have prevented the warring parties from employing biological weapons offensively if it had provided military advantages. This applies also to the use of bioweapons against civilian populations, where the Geneva Protocol is unclear.

While both sides' intelligence services undoubtedly enjoyed many successes during World War II, they failed to gather information on bioweapons. This activity was highly secretive in all states. This lack of coordination led to the warring states having only vague shadowy images of the opponent's bioweapon programme, while consistently exaggerating the threat posed by the enemy in this field. The result was that both the Axis powers and the Allies felt the necessity to develop biological weapons to respond to such attacks.[24,34]

The attitude towards the active use of bioweapons also varied among central figures, both civilian and military. Generally speaking, the perception was that such weapons should be reserved for retaliation if the enemy had already employed them. The lack of a clear emphasis on the offensive use of bioweapons was likely not due to ethical considerations. One exception may be the attitude of President Roosevelt's close friend and chief of staff, Admiral William Leahy, who stated in the summer of 1944 that 'to use microbes and toxins would violate any Christian ethics I have ever heard of and all known laws of war. It would be an attack on the enemy's non-combatant population. The reaction can be predicted – if we carry it out, the enemy will do it.'[34]

However, in the final phase of the war, the concept of total war that also affected civilian populations was accepted on both sides, as demonstrated by Germany's V1 and V2 rocket attacks on London, the Allied mass bombings of German cities, and the use of atomic bombs over Hiroshima and Nagasaki. Frederick Banting stated that 'it is equally effective to kill or incapacitate ten unarmed workers at home as to take a soldier out of action, and if this can be done with less risk, it would be advantageous to use all forms of warfare to achieve this'.

In response to claims that the use of bioweapons was unethical, Paul Fildes asked, 'Is it more moral to kill soldiers or civilians with high explosives than with bioweapons?'

The main reason biological weapons were not employed in the West during World War II is simply this: none of the warring powers were adequately prepared for it, aside from the anthrax cakes produced at Porton Down. Had effective bioweapons been available, it is not unlikely they would have been used if their military effectiveness had been anticipated.

Although a cold assessment of utility and advantages dominated thinking around biological warfare, it is tempting to ask: should not physicians heavily involved in the work with such weapons have had special reservations? Indeed, medical students at most universities no longer take the Hippocratic oath at the conclusion of their studies, but the principles of this oath have influenced Western medicine at all times since it was formulated over 2,500 years ago. This includes the principle that a doctor should not use their knowledge to kill or cause harm. Is it not then thought provoking that the Nobel Prize winner Frederick Banting – who through his discovery of insulin saved millions of lives – was now focused on developing deadly weapons against both military and civilians? In his diary, he fantasised about 'killing three or four million young Hun soldiers – without mercy – without feelings – It is our job to kill them.' Perhaps it is not without reason that mentions of Banting in reference works often either completely omit[39] or only touch briefly on his efforts as a pioneer in Canada's bioweapon programme.

Chapter 6

In the Shadow of Hiroshima – Bioweapons During the Cold War

In a speech in the small American town of Fulton, Missouri, on 5 March 1946 Winston Churchill declared that an iron curtain had descended across Europe. This was an acknowledgment that the relationship between the Western powers and the Soviet Union, their former ally, had become increasingly strained – an evolution that had begun even before the war ended. The Cold War was already under way, and it would last until the dissolution of the Soviet Union in 1991. This period was heavily marked by fears that the parties would resort to the use of nuclear weapons. This significantly dominated public debate, while a potential threat from bioweapons received limited attention. The reality, however, was that both main adversaries in the Cold War – the USA and the Soviet Union – were quietly building formidable bioweapons programmes. These programmes have only recently begun to come to light.

The USA's Targeted Investment in Bioweapons

In the last years of the war, the USA had focused intensely on developing biological weapons in close collaboration with its allies, the United Kingdom and Canada. By the war's end, the desire to continue the bioweapons programme remained strong. There had been hopes of obtaining interesting data from the Japanese programme through interviews with Ishii and his colleagues, but this turned out to be a disappointment. The Americans felt they had actually progressed further in this field than the Japanese. Nevertheless, further work was necessary, for even the war's intense efforts had not resolved all technical bioweapon-related problems.[34]

The centre for continued American work remained Camp Detrick (renamed Fort Detrick in 1956). Much of the fundamental research on relevant microbes was conducted here. The facility was expanded, including a massive spherical chamber – 'Eight Ball' – for experiments involving aerosol transmission of microbes, where both various test animals and humans were tested. Additionally, a ten-storey factory for the production of microbes was built in Arkansas, and a dedicated centre for field trials in outdoor settings was established in Utah.[34]

Officially, the USA's bioweapons programme was focused exclusively on defensive measures against such weapons that other states were feared to possess. The Soviet Union was particularly perceived as the threat. Authorities feared that the general public's attitude towards the use of bioweapons was negative, so activity at Fort Detrick was kept strictly secret. From the very beginning, it was clear that the operation would also aim at the development of bioweapons for offensive use. For several years, however, the stance was that the USA would only use biological weapons in retaliatory strikes if other states employed such weapons. This changed in 1956, and military documents show that at this time bioweapons were also envisioned as potentially useful for first-time use as a supplement to other weapons.

Several microbes were studied as potential weapon candidates, including bacteria that cause anthrax and tularaemia in humans. Certain viruses were also considered. Botulin toxin and some other toxins were part of the arsenal. Furthermore, work was done on microbes capable of destroying crops of rice, wheat, and rye.

Basic research achieved many important results – also of general interest to infectious medicine – and researchers were successful in identifying microbes that showed promise as bioweapons. However, the major challenge was determining the critical conditions for the practical use of these microbes in realistic military situations. How could effective doses be spread most effectively over large areas? What was the stability of the microbes under various climatic conditions? The Air Force, among others, was particularly interested in addressing these questions before it could consider using Fort Detrick's microbe candidates as bombs, which was the most immediate method of dispersing them.

Secret Bacterial Clouds Over the USA

Scientists at Fort Detrick concentrated in the early years on animal experiments in laboratories to determine infectious doses and other important factors. However, experiments under more realistic conditions were required before further progress could be made. This led to a long series of extensive experiments on humans, who were partly completely unaware of what was happening and, as we now know, faced significant risks. The experiments were carried out in complete secrecy. Knowledge of them came only after many years, leading to strong public reactions.

Many of these experiments were conducted in continued collaboration with the UK and Canada regarding bioweapons after the war. The first large field trials conducted in the open air were jointly performed with the British in the Caribbean from 1948 to 1949.[34, 41] The project, which was primarily administered by the British, was code-named Operation Harness. Bombs containing anthrax bacteria were detonated 20–50cm above cages of guinea pigs, sheep, and monkeys. The experiments were considered unsuccessful.

In the 1950s, the USA increased its investment in bioweapons, partly due to fears that the Soviet Union might use such weapons. There was heightened concern over the possibilities of spreading bioweapons over large areas, for example, by dispersing microbe-laden aerosols from planes or ships equipped with spray mechanisms. This new concept was called 'Large Area Coverage' (LAC) and was explored through a long series of experiments.[34]

From 1951 to 1954, a series of American cities were used secretly as testing laboratories to see how vulnerable urban communities were to bioweapons attacks.[34, 40, 42] Naturally, anthrax could not be used, so the experiments were instead conducted with two bacteria that were believed to be completely harmless to humans – *Serratia marcescens*, which produces a red pigment, and a distant relative of the anthrax bacterium, *Bacillus globigii*. During one of these experiments, a naval ship lay off San Francisco for a week, spraying an aerosol cloud of *Serratia marcescens* over San Francisco to observe the effect on a city with 800,000 inhabitants. An unexpected event occurred. A few days after the experiment, a small epidemic of infection with the supposedly harmless bacterium broke out at one of the city's hospitals, resulting in one death. The doctors considered the occurrence of infection with this allegedly benign bacterium so striking that the cases were published in a medical journal.

Subsequently, a committee was established to assess a potential connection between the cases of illness and the military experiments with *Serratia marcescens*. The committee did not reach a conclusion but recommended that the bacterium could still be used for such experiments.[42] This continued until 1968.

We now know that *Serratia marcescens* is not harmless at all, as it can cause serious, potentially fatal infections, particularly in people with compromised immune systems. These infections can be challenging to treat.

When an American newspaper first publicised this story in 1976, it led to strong reactions. There were Senate hearings that directed sharp criticism at the military authorities for allowing such experiments. Similar outbreaks of *Serratia* infections had also been observed elsewhere in the USA.

In hindsight, one might doubt whether the experiments in San Francisco in 1951 were the cause of the hospital epidemic, given that it turned out that the genetic type of the military's bacteria reportedly differed from the hospital bacteria.[42] Regardless, it is clear that these experiments posed a real risk to the population.[43]

Relatively recently, it has emerged that the other bacterium used in these experiments –*Bacillus globigii* – can also cause diseases in humans, including wound infections and intestinal infections, occasionally with fatal outcomes.[42]

Similar experiments were conducted against the cities of St Louis, Minneapolis, and Winnipeg, where the mentioned test bacteria were aerosolised over the cities from cars.

In several LAC experiments, the Americans also used the chemical substance zinc cadmium sulphide to study the dispersion of aerosols.[43] This substance is suitable as a trace maker since it fluoresces in ultraviolet light.

In 1951, bioweapon researchers conducted another experiment on unsuspecting individuals that should have raised red flags even then. This time, the fungus *Aspergillus fumigatus* was used as the supposedly harmless test microbe. This type of mould can cause some of the deadliest infections known, particularly in individuals with compromised immune systems.[13] As early as 1951, it was known that this fungus was pathogenic. Nevertheless, an experiment was launched using spores from *Aspergillus* at the naval base in Norfolk, Virginia. The base primarily employed African Americans, and it was desired to investigate whether people of colour were particularly susceptible to the fungus. Little is known about this experiment and any potential illness cases afterward. Here, an entirely unacceptable risk was taken. When the public learned about this experiment in 1980, it drew strong criticism – also due to potential racist aspects.[43]

In the 1960s, work on bioweapons intensified in the USA, partly due to the escalation of the Vietnam War. Researchers now systematically studied the dispersion of microbes from aircraft under various geographic conditions.[34] Experiments were conducted over jungle areas in Hawaii, and in Alaska to investigate the effects of biological weapons under Arctic conditions.

Additionally, tests were conducted with real bioweapons, not just test bacteria. On several small uninhabited coral islands near Hawaii, the effects on monkeys of bacteria that cause tularaemia and Q fever were studied, sprayed as aerosols from jet aircraft.

In new variants of bioweapon trials on unsuspecting citizens, aerosols with test bacteria were sprayed at both the main airport and the bus terminal in Washington in 1965. The following year, similar experiments were conducted in the New York subway.[34]

Researchers also wanted to test pathogenic microbes directly on trial subjects. They attempted to recruit volunteers for such trials, deemed safe since antibiotics were available against the bacteria. Some had the bright idea to use members of the Seventh-day Adventist religious minority. They were pacifists, but many served their military duty as medical personnel. Their religious leaders were told that the experiments would lead to effective protective measures against bioweapons, such as vaccines and new antibiotics. There was significant support for the trials, which took place from 1952 to 1969.[34]

The Adventists were infected with tularaemia, Q fever, and plague bacteria, among others. However, anthrax bacteria were also used in studies of skin infection, which is far less dangerous than lung infection with this bacterium. Antibiotics were available against all these bacteria. However, trials were also conducted with several viruses for which neither medications were available then nor are they today. Although these viral infections are typically asymptomatic or produce only mild illness symptoms, they can also cause severe, potentially fatal infections.

None of the over 2,000 trial participants died, but some fell ill for several weeks despite receiving antibiotic treatment. Late effects were reportedly not observed, but this is controversial.

This project, named Project Whitecoat, must be characterised as clearly ethically unacceptable. The participants were subjected to an obvious risk, which they were indeed informed of in advance. However, they were not informed that part of the purpose was the development of bioweapons for military use.

In the 1950s, the use of insects as carriers for microbes was evaluated.[5] As mentioned previously, the Japanese had already employed such methods in China and Manchuria, including plague-infested rat fleas. The Americans conducted a series of preliminary experiments to study the dispersal and survival of insects in bioweapons. In one of these experiments, code-named Big Itch, bombs containing rat fleas were dropped from aeroplanes. In another trial, bombs filled with the mosquito species *Aedes aegypti*, which can transmit yellow fever, were tested. Other dispersal methods for insects were also studied. Eventually, however, it was decided not to pursue insects in the bioweapons programme, as their effectiveness was considered relatively unpredictable.

From today's perspective, many of the experiments in the American bioweapons programme are questionable and somewhat shocking in their apparent lack of ethical considerations. Nevertheless, one should not forget that this occurred during the Cold War when the fear of a war with the Soviet Union was prevalent. In such a conflict, many military personnel – and even some politicians, including President John F. Kennedy – believed that all types of weapons could become relevant.

These experiments – which in no way provide a complete portrayal of the extensive US bioweapons programme up until around 1970 – indicate that the Americans were not only investing in their bio-defence but also in biological weapons for use in military conflicts around the globe – in temperate, tropical, and Arctic regions. During this period, a substantial arsenal of bioweapons was developed – primarily stored in various types of bombs, but possibly also in warheads on missiles.

President Nixon's 'Bombshell' in 1969

Bioweapon enthusiasts – scientists, military personnel, and certain politicians – were themselves struck by a 'bombshell'. In November 1969, President Richard Nixon declared that the USA would no longer invest in bioweapons but solely focus on defence against them. This declaration led to the BWC – Biological Weapons Convention – in 1972, which we will return to.[34]

At the time Nixon made his declaration, the USA had developed a sophisticated arsenal of microbes and toxins that were 'weaponised', meaning they were fully prepared for use as weapons against humans and plants. This included the bacteria that cause anthrax (spores from *Bacillus anthracis*), tularaemia (*Francisella tularensis*), brucellosis (*Brucella)*, and Q fever (*Coxiella burnetii*), as well as the Venezuelan equine encephalitis virus (VEE). Two toxins were also ready – botulin toxin and staphylococcal enterotoxin B. Additionally, three fungal species were 'weaponised' against plants: two species that attack wheat and rye, and one that destroys rice. There were significant stockpiles of all these microbes and toxins.

A range of bomb types and spray tank systems for dispersing microbes and toxins was also developed. In short, the USA was well prepared for the use of biological weapons in a military conflict. Therefore, Nixon's declaration was surprising and undoubtedly disappointing for many engaged in the bioweapons programme.

The background for Nixon's declaration was increasing pressure over time from scientist organisations, prominent individual researchers, and the media. Several arguments against bioweapons were presented.[44] Many believed that the use of biological weapons could lead to catastrophic consequences not only for humanity but also for the environment in terms of ecological damage. Nobel Prize winner Joshua Lederberg spoke of 'a virological Hiroshima'. These arguments were intertwined with the classical viewpoint of bioweapons as morally reprehensible.

A book by the well-known journalist Seymour Hersh (*Chemical and Biological Warfare – Americas Hidden Arsenal*), where he described the extensive bioweapons programme the USA had developed, caused considerable commotion.[45] In connection with the book, he wrote that 'human ingenuity could develop what would essentially be a "doomsday bug", a disease so uncontrollable that it would trigger a pandemic worldwide'. The strong criticism of the US use of chemical weapons during the Vietnam War also led to increased negative attention towards the American bioweapons programme.

Although many believed that bioweapons could be militarily useful, they asserted that the development of nuclear weapons rendered them redundant. Some also doubted the military utility of biological weapons, claiming they were untested and likely difficult to control, with unpredictable consequences.

As already noted during the Vietnam War, the USA had bioweapons ready for use. The prevailing view is that these weapons were never utilised. Is this the whole truth? As we will see later, strong claims have been made that the USA engaged in biological warfare on at least two occasions.

What actually happened to the American bioweapons programme after President Nixon's declaration? We will return to this.

The Bioweapons Programme in the UK – Fluctuating Enthusiasm

During World War II, the UK invested in both offensive and defensive bioweapon programmes. This support not only persisted but increased in the early post-war years.[34, 41] During this period, military leaders believed research in this field should be prioritised at a level comparable to that of nuclear weapons, and significant resources were allocated for it. This led to construction and expansions at the bioweapons centre in Porton Down. There, Paul Fildes' deputy during the war, Donald Henderson, had taken over as leader, while Fildes returned to his civilian research activities.[34]

Despite the increased resource availability, it became challenging to recruit researchers to Porton Down. Although the authorities had attempted to keep the bioweapons initiative secret during the war, critical comments in the media likely made research work in this field less attractive. While not all positions were filled, Porton Down did gain several qualified scientists.[34]

Intensive studies of suitable microbes for the bioweapons programme were conducted at Porton Down.[41] The primary method of dispersal for the weapons focused on bombs released from aircraft. Researchers at Porton Down collaborated with the Royal Air Force in a project code-named Red Admiral. As in the war years, there continued to be extensive collaboration with the USA and Canada, where several trials involving biobombs were conducted.

Researchers from Porton Down also participated in extensive outdoor experiments with microbes over maritime areas.[41] In this field, British researchers were pioneers in planning and execution. The first of these large-scale experiments was the aforementioned Operation Harness in Antigua in the Caribbean, conducted in collaboration with the USA.

In 1952, a similar trial – Operation Cauldron – was conducted off the coast of Scotland near the Isle of Lewis. Here, experiments involved aerosol clouds containing plague and *Brucella* bacteria. During a trial involving plague bacteria, a fishing boat passed through the bacterial cloud, and there was significant concern among the researchers that the fishermen would become infected with plague. The crew was never informed about the incident but was secretly monitored for more than three weeks for a potential plague outbreak, which ultimately did not occur. It was only after 50 years that the public learned about this.[46]

Operation Cauldron took place just a few kilometres from the Isle of Lewis, which had a population of several thousand. The scientists relied on the fact that there were usually offshore winds in the area, but if the wind had changed, thousands of islanders would likely have been infected with either plague or *Brucella* bacteria.

The following year, trials involving tularaemia and *Brucella* bacteria were conducted in the same location. In 1954, efforts were returned to the Caribbean, where experiments with the same bacteria, as well as the VVE virus, were conducted near the Bahamas. This proved to be the last of the large experiments over maritime areas in the British bioweapons programme.[41]

By the mid-1950s, enthusiasm for bioweapons had cooled significantly within British military circles – particularly in the Air Force, where there was no longer interest in developing a biobomb. A significant reason for this was that nuclear weapons were now considered far more critical, while there was still much uncertainty regarding the practical application of bioweapons.[34, 41]

At the same time, a new aspect of the threat from bioweapons emerged – the alarming possibility that an aircraft or ship equipped with spray mechanisms could disseminate aerosols with microbes that could spread like a cloud over large distances and affect a significant number of people, in contrast to biobombs, which would have a much more limited impact area. This required exploration – so-called LAC experimentation. In a way, this came as a boon for the researchers at Porton Down, who had feared reduced resources. It became necessary again to undertake extensive field experiments, where supposedly harmless bacteria or chemical tracers were aerosolised from planes, ships, or vehicles.

LAC exploration led to a series of experiments, both in the UK and the USA. The American experiments have been mentioned previously.

In the UK, these experiments were conducted between 1953 and 1964 using particles of the chemical substance zinc cadmium sulphide.[41] This substance was also, as mentioned earlier, used by the Americans in LAC experiments. In one of these experiments, 140kg of zinc cadmium sulphide was aerosolised from a plane for 40 miles over the North Sea. Samples taken from various locations in England and Wales indicated that nearly 30 million people would have received a dose of particles sufficient to cause infection with, for example, tularemia bacteria. A number of similar experiments were conducted in the following years, including over land. It has since been estimated that a total of 4,600kg of zinc cadmium sulphide was released as aerosol clouds from aeroplanes, ships, or cars.

In other similar LAC experiments, as in the USA, live, supposedly harmless bacteria –*Bacillus globigii* and *Serratia marcescens* – were aerosolised over large land areas. In some tests, the intestinal bacterium *E. coli* was also used.[41]

Bacillus globigii was also used in another type of experiment involving larger groups of people. To investigate the potential for using bioweapons in sabotage operations, a small box containing 30 grams of *Bacillus globigii* was released from a window of a train in the London underground in 1963. Air and dust samples were subsequently collected from the entire underground network over the next 12 days. It was found that the bacteria had spread several kilometres within the tunnel system and that trains passing through the aerosol cloud became heavily contaminated. Similar experiments were conducted in the underground system beneath government buildings in London.

A report from the British Ministry of Defence in 2001 describes a lengthy array of these experiments.[41] In reality, the entire country was utilised as a gigantic test laboratory where the inhabitants were unaware of what was going on. There has been extensive debate, which is likely not concluded, about the risks these experiments posed to the population. We now know that both *Bacillus globigii* and *Serratia marcescens* can cause infections, particularly in people with a weakened immune system.

There has also been disagreement concerning zinc cadmium sulphide and its associated risks. It has been claimed that this substance can cause cancers. However, an expert committee in the UK concluded that the doses involved in bioweapons testing were unlikely to be hazardous to health.

The aim of the extensive LAC experiments included mapping the significance of meteorological factors such as wind conditions, temperature, and solar radiation on the dispersion of microbes and their survival over time. Detection methods for microbes in airflows were also tested. Although these experiments and other research at Porton Down must retrospectively be described as ethically questionable, they contributed valuable new knowledge – not only concerning bioweapon issues but also for understanding aerosol transmission and respiratory infections in general.

From the late 1950s, it became clear that the UK would now only focus on bio-defence and no longer on bioweapons for offensive use. After the UK conducted its first successful atomic bomb test in 1952, nuclear weapons became the central focus. The UK was also one of the countries advocating for the work that culminated in the BWC in 1972 – the convention that prohibits the development, storage, and use of bioweapons.[34]

The Soviet Union Gradually Takes the Lead

The Russian research community had been impacted severely by the purges during the Great Terror of 1937 to 1938. In the years leading up to the end of World War II, there was evidently little dynamism in the Soviet bioweapons programme.[31,32] From 1945 onwards, however, there was a clear desire among leading military and political figures to continue the work on bioweapons and developing defences against such weapons. This had several causes.

The Russians had gained insight into the extensive Japanese bioweapons programme developed by figures like Ishii Shiro at Unit 731 in Ping Fan, Manchuria. When the Red Army entered Manchuria, they took control of Unit 731 and other bioweapons centres, gaining insight into some of the activities that had taken place there. More importantly, they received information from the 12 imprisoned Japanese scientists and officers who had been central to the bioweapons programme. As we have seen, the Japanese received surprisingly short prison sentences and were released between 1953 and 1956, likely because they provided extensive details about the Japanese biological weapons programme. This illustrated to the Russians the potential of bioweapons. The Japanese research results were also of interest to the Russians and may have spurred further activity.[30,34]

Moreover, the Russians were convinced that the Western powers, primarily the USA, were investing in bioweapons. They were aware of the close cooperation in this field between the USA, the UK, and Canada during the war. They knew that the Americans had secretly interviewed Ishii and his colleagues, thereby acquiring valuable Japanese data. Therefore, it was believed that the Soviet Union had to invest in both defences against and the development of bioweapons to be able to respond to attacks with such weapons. The Cold War was already under way.[31]

The Dark Years

The Soviet Union was a totalitarian and highly secretive society. It is not surprising, therefore, that the West knew very little about the Russian bioweapons programme, which was secretive even within the Soviet Union. I have termed the first period 'the dark years', lasting until around 1970, when a shift occurred in the dynamics of the Russian bioweapons programme.[31,47]

During these dark years, we know that there were several important military bioweapons centres in the Soviet Union, including the Kirov centre. Eventually, a large facility was established in Sverdlovsk (today Yekaterinburg) – which we will hear more about – and an important centre in Zagorsk. There were also several other military-run bioweapons centres. However, there were also several civilian institutions involved in the field, including the Ministry of Agriculture, the Ministry of Internal Affairs, and the Russian Academy of Sciences.

It is reasonable to assume that the Russian bioweapons programme of the 1960s made little progress and was almost stagnant. One likely reason for this was the destructive impact that Trofim Lysenko had on genetic research in the Soviet Union for many years. Lysenko was a biologist and agronomist who wholly rejected modern genetics, which was based on the ideas of Gregor Mendel and Charles Darwin. Instead, he relied on fairly primitive notions that included the belief that acquired traits could be inherited. This was supported by poor research and wishful thinking.[48, 49]

From the late 1930s, Lysenko gained increasing influence in the Russian research community, primarily because Soviet leaders – particularly Joseph Stalin – supported him, as his ideas were politically useful. Lysenko attained top positions in the research world. As the 1940s progressed, Lysenko's ideology became entirely dominant, and any deviation from it was punished severely. Many researchers lost their positions, were imprisoned, and even executed.

When Lysenko's powerful protector, Stalin, died in 1953, Lysenko lost much of his influence, and after 1964 he was completely sidelined. Modern genetics regained a foothold in the Soviet Union.

However, Lysenko's grip on parts of Russian research had caused genetic research to lag behind Western countries. This had also adversely affected the fundamental research in the bioweapons programme. Regardless, the military leadership around 1970 considered shutting down operations. President Nixon had already declared in 1969, as we recall, that the USA would no longer have offensive bioweapons. Nevertheless, a closer evaluation culminated in the exact opposite in the Soviet Union: they decided to increase their focus.

By this time, the Soviet Union had likely prepared several microbes as weapons – both bacteria that cause plague, tularaemia, Q fever, and typhus, as well as viruses, including the smallpox virus.[47] Earlier in the period, a significant amount of effort had also been placed on botulin toxin, which was gradually set aside as it was deemed less suitable for practical use as a biological weapon.

Biopreparat – A Giant Company in the Service of Death

The revitalisation from the early 1970s led to the Soviet bioweapons programme entering a very active and fruitful growth period. This resulted in the Soviet Union, by the time of its dissolution, having the most extensive and sophisticated bioweapons programme the world had ever seen. We know quite a bit about what transpired during these years thanks to three Russian defectors, including the previously mentioned Ken Alibek, who had first-hand insight into the Soviet bioweapons situation.[33] Interviews with Russians after the fall of the Soviet Union have also provided important information. Furthermore, the West has gained access to some Russian archives. However, significant gaps remain in our knowledge regarding the development of biological weapons in the Soviet Union.

What caused the authorities to choose to increase the investment in the bioweapons programme from the early 1970s instead of shutting it down? Even before President Nixon's groundbreaking declaration regarding biological weapons in 1969, the Soviet Union had been one of the driving forces behind a new international treaty to abolish bioweapons that led to the BWC in 1972. Thus, the Soviet Union chose from the outset to violate this convention, which it had advocated! A single person seems to have played a central role in this important decision – Yuri A. Ovchinnikov.[31, 47]

The biochemist Ovchinnikov was an outstanding scientist who, in 1970 – aged only 36 – was highly respected among Soviet researchers and had already become the head of one of the Academy of Sciences' research institutes. He was ambitious and an active member of the Communist Party, which was necessary to achieve top positions in the Soviet research community. He was a skilled organiser and network builder. Eventually, he obtained pivotal positions in both the research world and the Communist Party and was reportedly confided in by Leonid Brezhnev – the powerful General Secretary of the Communist Party.

Ovchinnikov recognised that drastic measures were necessary for the Soviet Union to catch up with the West's advancements in molecular biology and biotechnology. To convince the Soviet leadership, he argued that bioweapons development was critical for national security and that modern biotechnology was an entirely necessary foundation for this work. He was heeded, and the Soviet leadership decided to invest in biotechnology. Ovchinnikov himself received significant resources for a massive expansion of his research institute, which other researchers referred to as the 'Taj Mahal of Russian biotechnology'.

Furthermore, the Soviet leadership decided on a large-scale restructuring and expansion of the Soviet bioweapons programme. Following the advice of figures like Ovchinnikov, large portions of the programme were centred on a network of civilian research laboratories across the Soviet Union, where operations were officially claimed to have merely civilian purposes. This network was named Biopreparat.[31, 47]

Biopreparat was, in reality, under the control of the Ministry of Defence. For many years, the head of Biopreparat was the medically trained general Yefim Smirnov. Smirnov, who held a leading position in the Red Army's medical services during the war, was also previously central in the Russian bioweapons programme. He had been Stalin's confidant until he fell from grace just months before the dictator's death in 1953. This was due to Stalin's paranoid reactions to the 'doctor's plot', which was allegedly a threat to the Soviet leader.

Smirnov was the most ardent advocate among military leaders for the necessity of biological weapons. He collaborated closely with Ovchinnikov, and together they likely share the dubious honour of establishing the world's most extensive bioweapons programme.

Ovchinnikov is not the only researcher who chose to use their scientific insights to develop bioweapons. We have mentioned previously Nobel Prize winner Frederick Banting in Canada, Auguste Trillat in France, and Paul Fildes in the UK. It is likely that Ovchinnikov was not primarily focused on bioweapons but rather aimed to strengthen modern molecular biology and biotechnology for peaceful purposes in the Soviet Union. As he is said to have remarked: 'If we in the Central Committee offer ten medicines, no one will support us. No one will give us money for medicines. But if you offer one weapon, you will get full support.'[47]

In addition to numerous research laboratories, Biopreparat also had several centres for producing microbes for the bioweapons programme, including the previously mentioned centre in Sverdlovsk. Furthermore, the civilian research laboratories collaborated with the military, which remained operational.

Ambitious and Deadly Goals

Biopreparat had very ambitious and extensive research objectives in its efforts to develop promising microbes as weapons.[31, 32, 47]

First, they started from the microbes with which they already had experience and aimed to alter their properties to make them even more dangerous. They sought to increase both transmissibility and virulence –

disease-causing abilities – as well as to boost bacterias resistance to antibiotics. They also manipulated microbes to render them resistant to vaccines.

Additionally, the researchers at Biopreparat were heavily focused on the previously mentioned emerging infectious diseases – that is, newly discovered infectious diseases – that have appeared in increasing numbers since the mid-twentieth century. This included several viruses that cause so-called haemorrhagic fevers, such as the Marburg, Ebola, Machupo, and Lassa viruses.[13] Here, the Russians saw promising candidates for deadly weapons. Samples of these newly discovered viruses were partly acquired through 'legal' means from laboratories in other countries. The Soviet intelligence service, which after Beria's fall in 1953 was called the KGB, was represented in the leadership of Biopreparat. The KGB was also internationally active in procuring promising viruses for the bioweapons programme using more shadowy means. Among the mentioned new viruses, particular interest was given to the frightening Marburg virus – a relative of Ebola. This virus has a lethality rate of up to 70%.

Numerous more or less authentic stories circulated about the KGB's eagerness to acquire new viruses. KGB agents reportedly secretly dug up the bodies of the first individuals who died from infection with the Marburg virus in Europe due to a laboratory outbreak in the German city of Marburg in 1967.[13]

However, the scientists did not settle for focusing solely on existing microbes. They employed new genetic technologies in attempts to create entirely new microbes – so-called chimaeras. These chimaeras were assembled from specific genetic elements from various microbes in order to make them more dangerous in different ways. Among other things, they worked on 'fusing' the smallpox and Ebola viruses in this manner. It remains unclear how far they progressed in this work.

Perhaps the most innovative attempts, which bear the hallmarks of science fiction, were conducted in the top-secret Project Vektor programme.[50] In one of these projects, efforts were made to develop a bioweapon that causes a form of multiple sclerosis. They took the gene for myelin basic protein – the insulating material for nerves in the brain and spinal cord – and inserted it into a bacterium, which subsequently produced the protein. When guinea pigs were infected with this bacterium, the animals developed paralysis after two weeks and died due to an autoimmune reaction against the myelin protein produced by the bacteria.

In other experiments, they introduced genes that control the production of neurotransmitter molecules into bacteria – molecules that convey signals between the nerve cells in the brain. When such genetically modified bacteria infect, the overproduction of these molecules will disrupt brain functions in various ways, altering the thought processes, behaviour, and emotions of the victim.[50]

Fortunately, the attempts in Project Vektor never led to finished bioweapons – at least not before the dissolution of the Soviet Union.

Hell Island in the Aral Sea

The microbes chosen as promising weapon candidates naturally needed to be tested in practice. Essentially, the Soviet programme focused on dispersing microbes as aerosols. What we know with a reasonable degree of certainty is that it was primarily bombs and grenades that were attempted as carriers for the microbes. Some Russian defectors have claimed that intercontinental missiles were also equipped with bioweapons in their warheads. However, this remains uncertain.

What is clear, however, is that extensive, realistic field trials of bombs and grenades containing microbes were conducted. Like everything else concerning the Soviet bioweapons programme, such experiments had to be executed in utmost secrecy. Here, situated in the vast Soviet empire, there was an ideal location 'in the middle of nowhere' for such top-secret trials.

In the middle of what was then the world's fourth-largest lake – the Aral Sea on the borders of Kazakhstan and Uzbekistan – lay Vozrozhdeniya Island. The name has been translated as 'rebirth' or 'resurrection', referring to the birth of the Soviet Union. Due to the trials on the island, it acquired a less poetic name among many – 'Hell Island'.[31, 47]

Vozrozhdeniya Island had been used for bioweapons trials since the 1930s. However, the facility that came to be known as Aralsk-7 began modernising and expanding significantly in the 1950s. Researchers and soldiers lived on the island, and laboratories and animal quarters were constructed. The main headquarters were located in the small village of Aralsk on the northern shore of the Aral Sea.

In the animal quarters, there were various test animals, both large and small. However, it was primarily baboons imported from Africa that were used in extensive outdoor experiments with bombs, grenades, and spray devices. Up to one hundred baboons, tied to stakes, were used for the experiments. There have also been claims that human trials were conducted here, but this remains uncertain.

The entire spectrum of microbes that were particularly focused on was tested on Vozrozhdeniya Island – both bacteria and viruses. It is not known for certain whether serious accidents occurred involving infections among those working there. What is known, however, is that the experiments led to an outbreak of smallpox in the island's vicinity.[51] In the summer of 1971, a research ship from the Russian Navy was on an expedition in the Aral Sea to study the ecological changes that had caused partial drying out of the sea. A woman on board became infected with the smallpox virus that had spread through the air from one of the trials on Hell Island. She recovered but carried the virus to the local population in Aralsk. In total, ten individuals were infected, and three died. The health authorities managed to halt the epidemic through aggressive quarantine measures and smallpox vaccinations. The Soviet authorities never provided an explanation for the outbreak of smallpox, and the victims never received any compensation.

Although those working on the island did not likely suffer from infections, the facility was regarded as a rather unpleasant workplace. It was extremely isolated, and the soldiers on the island were given such poor rations that they stole food from the baboons.

With the official dissolution of the Soviet Union in 1991, the bioweapons facility on Vozrozhdeniya Island was also shut down. Later inspections revealed that there were still large quantities of anthrax spores in the soil. This was due not only to the anthrax experiments but also because the Soviet authorities had buried substantial amounts of their anthrax stockpiles at the end of the 1980s.

The outbreak of smallpox as a result of the activity on Vozrozhdeniya Island was relatively limited and effectively hushed up. A few years later, however, another outbreak occurred due to an accident in the Soviet bioweapons programme. This would have considerably more serious consequences.

The Anthrax Outbreak in Sverdlovsk

At the previously mentioned bioweapons facility in Sverdlovsk in the Ural Mountains, mass production of anthrax spores was ongoing within the Biopreparat system. On 2 April 1979, the facility released a cloud of anthrax spores that followed the wind direction southward. During the night, routine maintenance had been performed on the filters in the ventilation system that discharged air from an anthrax production room. The filters normally removed anthrax spores from the outgoing air. Due to an oversight, maintenance workers forgot to replace a used filter with a new one – and the result was that anthrax spores streamed out into the open.[28]

After a couple of days, several people near the facility began to exhibit symptoms. Initially resembling a common respiratory infection, these symptoms quickly transitioned to more severe illness characterised by high fever, lung symptoms, dark rashes, and other signs typical of anthrax. More and more patients were admitted to hospitals, eventually leading to a string of deaths. To this day, there are no reliable numbers of how many died during the outbreak. Estimates that include both the civilian population and employees at the facility range from under 100 to 600.[28, 52]

Operations at the bioweapons facility, like everything else in the Biopreparat system, were strictly secret. The population of Sverdlovsk was unaware of what was occurring there as the facility's management did not disclose any information about what had happened. Local doctors were not familiar with the symptoms of anthrax, and it took some time before it was recognised. At that point, they were bewildered regarding the origins of the outbreak as there had not been any naturally occurring anthrax in the area for many years.

The leaders at Biopreparat initiated an extensive cover-up operation. A statement was released claiming that the outbreak resulted from individuals consuming anthrax-infected meat. The population was warned against eating food from stray animals, and a façade hunt was initiated against such animals. Mass vaccinations against anthrax in the area were also carried out.

Rumours about the outbreak eventually reached beyond the borders of the Soviet Union. Western intelligence agencies quickly obtained information, but initially this was suppressed by politicians who wished to believe that the Russians were adhering to the BWC. It was not until half a year later that the Western press managed to obtain the news. When the CIA received reliable information about what had transpired, President Jimmy Carter's administration interpreted this unsurprisingly as evidence that the Soviet Union was producing bioweapons in violation of the BWC that it had ratified in 1975. When the USA issued a sharp demand for an explanation from the Soviet Union, the response was a total denial. This was followed by the false explanation that the outbreak resulted from infected meat. A special delegation from the Soviet Union travelled to the USA to present this explanation. Initially, it was accepted by leading bioweapons experts in the USA.[28, 52]

When, in 1992, detailed information about the cases of illness in Sverdlovsk was disclosed, it became clear that the Soviet explanation was deceitful. The symptoms indicated that the cases involved anthrax where the spores had entered via the respiratory tract – not through the ingestion of food, which would present completely different disease profiles. Respiratory infection, as seen in the Sverdlovsk epidemic, is rarely observed in natural anthrax outbreaks.

In 1992, Boris Yeltsin, then President of Russia, also admitted that the anthrax outbreak was due to a release from the bioweapons facility. As far as is known, no compensation was ever paid to the relatives of the victims.[34]

The investigation into the Sverdlovsk incident has yielded several useful insights. It can be viewed as a realistic illustration of how a bioweapon attack unfolds and underscores the importance of rapid diagnosis and early treatment. It also provided new and significant insights into the symptoms of anthrax and demonstrated that the infection has the most severe course in older individuals.

Finally, the outbreak illustrated the risks of leaks from laboratories working with dangerous microbes – not only in terms of bioweapon contexts. This risk is today more relevant than ever, especially in connection with the ongoing debate regarding so-called gain-of-function research. We will return to this later.

The Bioweapons Programme at the Dissolution of the Soviet Union

We have received information from several civilian defectors with significant insights into the Soviet bioweapons programme and there were a range of interviews with Biopreparat employees after the dissolution of the Soviet Union. Nevertheless, there is still much we do not know about the practical results of the programme. We have no concrete knowledge regarding the Red Army's views on the use of biological weapons in warfare. No central figure in the military segment of the bioweapons programme has defected or agreed to be interviewed. According to several civilian scientists in Biopreparat, however, there were plans for the use of bioweapons in various military scenarios, combined with both conventional and nuclear weapons.[47]

Although Russia, in 1992, guaranteed full transparency into its facilities through an agreement with the USA and the UK, this never occurred. Nonetheless, we can assert with some certainty how far bioweapons development had progressed by the time of the Soviet Union's dissolution.[47] What can be characterised as the Soviet Union's second-generation bioweapons had, to a large extent, built upon the microbes researched previously, albeit in improved forms. Some new microbes had also emerged, especially in the virus group. Some of these bioweapons were intended for humans, while others targeted animals and plants.

The exact number and types of microbes in the programme remain uncertain. Among bacteria, there were certainly plague, Brucella, tularaemia, and glanders and pseudoglanders on the list, in addition to those causing Q fever (*Coxiella burnetii*) and Legionnaires' disease (*Legionella pneumophila*). Botulin toxin is also included. In the virus group, there were both those that attack humans, such as smallpox and Marburg, and a series intended for use against various forms of livestock, including cattle and pigs (Rinderpest, foot-and-mouth disease, and swine flu). Bioweapons targeting a range of essential grain crops, including corn and rice, were also developed.

At a number of the Biopreparat facilities, there were permanent stockpiles of the relevant microbes ready for filling in bombs and spray tanks. Another type of facility was designated for the mass production of bioweapons upon mobilisation. These were located near airports and railway stations, and capable of producing several hundred tons of relevant microbes over a year. Several facilities, for instance, could produce at least 300 tons per year of anthrax, tularaemia, and plague bacteria.

The dispersion of microbes occurred as aerosols in two ways – either with bombs or with spray systems from aircraft types specifically suited for the respective method. This involved 'cluster bombs' consisting of a larger number of small bombs. The spray systems were deemed most suitable for dispersing microbes over wider areas.[31, 47]

The military plans for the utilisation of the bioweapons remain unknown. However, it is reasonable to assume that the Russians did not intend to use them tactically, that is, on the battlefield against enemy forces. Instead, such weapons were considered suitable for use behind enemy lines to incapacitate vital functional areas like airports, ports, and depots. They could also be employed strategically over large portions of enemy territory to paralyse both civilian and military resistance.[47]

Although the Soviet Union likely had a fairly advanced bioweapons programme, there are no solid indications that such weapons were ever used. Some have claimed that there is a possible exception – the highly publicised 'Yellow Rain' affair, which will be discussed later.

In 1992, Boris Yeltsin admitted that the Soviet Union had maintained an extensive bioweapons programme and that this would now be dismantled. We will return to the developments following the end of the Cold War.

Dirty War in Rhodesia

The Cold War, which characterised the relationship between the superpowers, coincided with the decolonisation of Africa and Asia. More and more former colonies became independent states. In southern Africa, Zambia and Malawi gained independence in 1964. However, in the neighbouring country of Southern Rhodesia – today's Zimbabwe – the ruling white minority population did not want independence with a black majority government. When the far-right party Rhodesian Front won the election in 1965, the new government, led by former RAF pilot Ian Smith, declared the country independent under the name of Rhodesia. Smith's regime was condemned by the UN and was not recognised by any state. Under Smith, the apartheid policies were further tightened, and previous attempts at liberalisation were reversed.[53]

Since white settlers from South Africa invaded Zimbabwe in 1890, there had been strong resistance to the white minority rule in the country. This ended with a large-scale uprising at the end of the nineteenth century, which was suppressed by British and South African forces. In the 1950s and '60s, several nationalist parties were established by the black majority population fighting for majority rule. Two parties eventually became dominant in the liberation struggle: Zapu, supported and financed by the Soviet Union, and Zanu, which received support from communist China. Both parties had a military wing that, from the mid-1960s, began guerrilla attacks against the white regime from bases in neighbouring countries.[28, 54]

Initially, Smith's regime was able to suppress the guerrilla attacks, but from 1972, the war escalated again. The rebels gained significant advantages when Portugal withdrew from neighbouring Mozambique in 1975, where the new Marxist regime now supported the guerrilla forces, enabling them to establish camps in the country. The Rhodesian government had specially trained security forces that were deployed against the insurgents, including native units known as the Selous Scouts, who infiltrated rebel areas under the guise of being guerrilla soldiers and mapped out their forces. Many of these operatives were former insurgents who, during captivity, had served the whites and could therefore no longer return to their own villages. With their covert operations in guerrilla areas, they could inform regular Rhodesian military units about enemy positions and strengths, enabling them to move in to eliminate the guerrilla fighters. This was a highly effective strategy.

However, new guerrilla forces continually flowed across the borders. It became clear to the government, after intense calls from both the white population and the security forces, that a more forceful effort against the insurgents was necessary. Aside from substantial support from South Africa, Rhodesia was fairly isolated both in the region and in the world at large. International sanctions had also been imposed against the regime. It was at this time that Ian Smith's government chose to employ 'dirty tricks' – biological and chemical weapons.

The leadership of the security forces contacted doctors and chemists at the University of Rhodesia in Salisbury (now Harare) and asked for input on the development of biological and chemical weapons. Robert Symington, a professor of anatomy at the university, became the research leader for this programme, working with a relatively unprofessional group of students.[53]

The group focused on the biotoxin ricin, in addition to the poisons thallium and parathion, and the anticoagulant warfarin, which induces bleeding. Ricin was primarily used as an assassination weapon against individuals, such as leading opposition politicians. Some have claimed that the security forces tested various poisons on imprisoned guerrilla soldiers at a secret, remote centre, but this is not well documented.

Initially, thallium, which is commonly used as rat poison, was used in attempts at mass poisoning of guerrilla soldiers. The poison was injected into meat products that Rhodesian agents ensured the enemy received. On one occasion, guerrilla soldiers provided such meat to starving villagers, resulting in several fatalities.

There are reports that cholera bacteria were part of Rhodesia's biological weapon arsenal. This microbe weapon was particularly used in the border areas against Mozambique and inside that country near the guerrilla forces' camps, leading to several cholera outbreaks. However, the Rhodesians feared that cholera could spread into their own country – and when the tactic proved not particularly effective, it was discontinued.

There are many both fascinating and disturbing aspects of Rhodesia's 'dirty war'. One of the really big questions is whether Ian Smith's government was responsible for the anthrax epidemic that broke out in parts of Rhodesia in 1978. By this time, it had become clear to many in the white regime that victory in the fight against the rebels was not possible. The regime was, as mentioned, backed against the wall, without international support aside from South Africa. At the same time, the bitterness in the conflict further escalated when a civilian Rhodesian passenger plane was shot down by ZAPU insurgents with a Soviet-produced missile, and the surviving passengers were massacred.

The anthrax epidemic that broke out in late 1978 was the largest the world had seen in over 200 years.[53] According to the latest analysis of the epidemic as it developed from January 1979 to December 1980, over 170,000 cattle were infected with anthrax, while over 17,000 people were affected, resulting in around 200 deaths.[53]

The common belief is that the anthrax bacteria originally came from South Africa and has since been exported to other parts of the world over the centuries. In southern Africa, anthrax ravaged large cattle herds until the Europeans began extensive vaccination of cattle against anthrax from 1920, leading to a drastic reduction in infections. This also applied to Rhodesia, where anthrax had become a rare infection in the years prior to 1978. By this time, doctors in Rhodesia had hardly seen the infection in humans.[53]

Did the outbreak of anthrax in 1978 have natural causes? Or was the epidemic a consequence of anthrax being used as a biological weapon by Rhodesia's white regime?

The claim that the epidemic was man-made was first made by Meryl Nass, an American doctor who was in Rhodesia at the time. Nass, an expert on biological weapons, reacted to the fact that this anthrax epidemic had several striking features that are not usually seen in naturally occurring outbreaks.

The scale of the epidemic was, as already pointed out, unusually extensive. There was also no single starting point for the epidemic, which seemed to have started simultaneously in various locations. Furthermore, the areas particularly affected were, to some extent, geographically separated, which did not suggest a natural spread. It was also striking that the epidemic mainly struck the so-called Tribal Trust Lands that were allocated to the indigenous population, while the areas where white farmers were engaged in extensive commercial agriculture were almost unaffected. In several of the affected areas, anthrax had not been seen for years, and there were no obvious natural explanations that could account for the outbreak, such as specific climatic conditions. At the same time, rumours circulated that the Rhodesian government was using chemical and biological weapons in the fight against the insurgents. It was not only Nass who reacted to the striking features of this epidemic.

In recent years, some of the former representatives of the security forces have claimed in interviews that the regime indeed used anthrax as a biological weapon in the Tribal Trust Lands. The purpose of this strategy could be twofold. First, they would target the meat production in the affected areas that the insurgent forces depended on. Secondly, the anthrax outbreak could create the impression that it had been brought into Rhodesia from Mozambique by the insurgents, thereby weakening the support of the black villagers for the guerrilla forces.[28]

No one has been able to explain how the anthrax was potentially used practically as a biological weapon. The most likely scenario would be the use of 'anthrax cakes' similar to what the British produced during World War II.

There is little doubt that the white regime in Rhodesia used both biological and chemical weapons in the struggle with the insurgent forces, even though the use of biological weapons was quite unsophisticated. Anthrax may also have been used, but there is no written documentation for this from the relevant period. It is also somewhat striking that the black majority regime under Robert Mugabe – leader of ZANU – which took over after Ian Smith in 1980, has never initiated an investigation into this. Meryl Nass, who first proposed the biological weapon hypothesis, had several suggestions for further investigations to test it, but these were evidently not followed up. Later authors writing about this have largely built on Nass's observations.

Overall, there are many indications that the extensive anthrax epidemic of 1978 was triggered by the white regime's use of biological weapons, however, this cannot be said to be definitively documented.[55] Recent research has not ruled out that natural causes may have been decisive. In the years following this major epidemic, there have also been new outbreaks of anthrax in Zimbabwe.

There are grounds to believe that the support Ian Smith's regime received from South Africa also included assistance with the regime's use of biological and chemical weapons. There were said to be a significant number of agents from South Africa in Rhodesia who assisted during the regime's 'dirty war'. It is possible that South Africa actually used Rhodesia as a 'testing laboratory' when planning its own far more extensive biological weapons programme, code-named Project Coast. After the fall of the Smith regime, many whites who had been involved in Rhodesia's programme emigrated to South Africa, where they may have continued their shadowy activities.[56]

'Dr Death' and South Africa's Bioweapons Programme

Ian Smith's regime lost power in Rhodesia, and the country became free in 1980 as Zimbabwe. The Republic of South Africa remained the last remaining stronghold in Africa for white minority rule. After the country exited the British Commonwealth, the so-called apartheid policy gradually developed as a protection for the privileged position of the white minority. Racial policies led to South Africa becoming increasingly isolated in the world, and international sanctions against the country were enacted. After neighbouring countries were decolonised, South Africa entered military conflict with liberation forces in Mozambique, Angola, and South-West

Africa (now Namibia). At the same time, internal resistance to the white minority regime, led by the Afrikaner-dominated National Party, increased. The African National Congress (ANC) became central in the opposition from the black population.

In 1978, Pieter Botha from the National Party became president. He viewed the conflict between whites and blacks in South Africa through the lens of the Cold War. According to Botha, South Africa was the most important bastion on the African continent against Soviet international advances, and he viewed both the black opposition in South Africa and the liberation forces in neighbouring countries as Soviet-backed troops in the struggle. In his speeches, he emphasized that his country was under a 'total onslaught' that had to be met with a 'total strategy', which meant that South Africa now had to employ all means in the fight.[56]

Project Coast Established

This was the background for South Africa's extensive bioweapons programme, code-named Project Coast, which was established in 1981.[56, 57] The same programme also included the development of chemical weapons. According to official documents, the justification for the entire programme was that South African forces needed to develop defences against chemical weapons allegedly used in the fighting in Angola, where Cuban forces participated. In reality, there was a desire for an offensive programme to make both biological and chemical weapons available as part of the 'total strategy'. This has since become clear. From the outset, the plans were therefore in violation of the Biological Weapons Convention (BWC) that South Africa had ratified.

The young military doctor Wouter Basson – a heart specialist and the prime minister's doctor – was tasked in 1981 with conducting a comprehensive overseas trip to study various countries' programmes for biological and chemical weapons. He returned after a few months with a report and his recommendations for a similar programme in South Africa. Basson was a lieutenant colonel in the South African Medical Service (SAMS), the medical branch of South Africa's military forces, which was on par with the army, air force, and navy. Project Coast was now linked to a special division of SAMS that was officially tasked with protecting the country against attacks using biological and chemical weapons.

We have seen several examples from other states where the establishment of bioweapons programmes has largely depended on the efforts of individuals. In the case of South Africa, there is no doubt that Wouter Basson was the father of the programme and also later the key figure in the work to develop biological and chemical weapons.

The Project Coast programme that was now initiated would not only develop and test biological and chemical weapons but also facilitate industrial production of such agents. Furthermore, the programme was to assist the military security forces in the application of these agents against the state's enemies both within and outside its borders. Additionally, documents reveal that Project Coast was also to conduct its own operations using biological weapons and chemical agents. Thus, this programme was very much offensive in nature.

Project Coast was strictly secret. The programme was organised so that the activities could not in any way be connected with South Africa's military forces. Several facilities were therefore established as 'front companies' that ostensibly conducted purely civil activities. The fundamental research on bioweapons occurred at the Roodeplaat Research Laboratories (RRL). At another laboratory – Delta G Scientific – chemical weapons production took place. A third 'front company' – Infladel – handled administration and financial transactions. In reality, all these companies were military funded.[56,57]

The use of ostensibly civilian companies and facilities as a cover for bioweapons activities was Basson's clever idea when he returned from his 'study trip'. We have encountered this arrangement before in the form of the Soviet Biopreparat organisation. Since it is known that Basson also had contacts there and even visited Moscow, he may have drawn inspiration from the Soviet model.

There were no problems recruiting researchers for Project Coast.[57] Basson personally started the recruitment from South Africa's university environments, selecting individuals he believed had relevant expertise. Potential recruits for the programme were offered high salaries and the opportunity to continue their own research projects. This attracted many, as research opportunities in South Africa were relatively limited. Several of the researchers also felt that the country was in danger, and that they were fulfilling their patriotic duty by joining the programme. Over time, the employed researchers also contributed to further recruiting among their colleagues.

It was unlikely to have been clear to all the newly hired scientists that the work involved the development and active use of biological weapons. Gradually, this became more evident, but there are not many documented examples of scientists withdrawing from the programme for ethical reasons. Also, much of the operations within Project Coast were kept secret, so the researchers often did not know what their results and products were being used for. Some researchers claimed later that they were never aware they were working for the military.

Initially, Project Coast was to be overseen by the head of military health together with a committee. In reality, it was only Basson who had an overview of the activities. He received unofficial input and requests from certain military top leaders.

At the laboratories at RRL, work was done with a number of microbes and toxins. It is fairly certain that the bacteria causing anthrax, cholera, plague, brucellosis, 'mouse typhus' (*Salmonella typhimurium*), and gas gangrene (*Clostridium perfringens*) were considered to be interesting weapons. Among toxins, there was a focus on botulin toxin and ricin, and the venom from the black mamba – one of the most poisonous snakes in Africa – was also studied. It has also been claimed that Basson and his colleagues acquired the deadly viruses Ebola and Marburg.[56, 57]

The original background for Project Coast was reportedly military, where biological and chemical weapons were to be employed in the fight against hostile forces and insurgents. However, there is little evidence that this actually happened on a larger scale. On one occasion, cholera bacteria were introduced into a water reservoir for a guerrilla force in Namibia, but with highly dubious results as the water had been chlorinated, neutralising the bacteria. On another occasion, sugar contaminated with the bacterium *Salmonella typhimurium* was smuggled into an ANC meeting. Whether some of the participants suffered from diarrhoea remains uncertain. Nevertheless, the attack had no political or military consequences.

The reason for the lack of military success was likely primarily that Basson and his colleagues never solved the critical problem of effective dissemination, which is best accomplished with aerosols, of which we have seen numerous examples. Only a few initial attempts were made in this area, therefore they were left with fairly primitive and ineffective means of dissemination.[56]

Assassinations as Routine

Biological and chemical weapons were used by Project Coast in one of the most extensive assassination programmes against political opponents that the world has ever seen.[28, 56] The inventiveness was impressive. Many may remember from the classic James Bond films that Bond, before a new mission, is offered various refined technical weapons from the technology specialist, Q. I believe Bond might have nodded in approval at the many imaginative murder weapons Basson and his people could offer agents in the field. Among Basson's possessions was found a list of such weapons that had been distributed to agents over a seven-month period – cigarettes with filters infused with anthrax, anthrax on chewing gum and chocolate, botulin toxin in beer bottles, Salmonella bacteria in whisky and deodorants, along with various toxins used in equally imaginative ways. One specialty, possibly imported from Rhodesia, was underwear infused with the dangerous nerve toxin parathion, which can penetrate intact skin. Other specialties included umbrellas and walking sticks that could be used to inject toxins such as ricin.

These assassination methods were employed against political opponents both in South Africa and in other countries, including against ANC representatives in London. We will likely never know how many fell victim to Project Coast's numerous biological and chemical weapons. Who can definitively determine that a death due to an infection was caused by bioweapons and not natural causes? However, there are certain examples that illustrate the ruthless brutality exhibited by South Africa's security forces in the fight against the country's opponents. An agent of the security forces confessed to having executed several hundred imprisoned soldiers from the Namibian liberation movement SWAPO by injecting muscle relaxants that led to paralysis of the respiratory muscles. The bodies were then thrown into the sea from helicopters.[56]

Also Racial Hygiene Objectives

Under the auspices of Project Coast, large quantities of narcotic substances were also produced, including the popular drugs methaqualone (Quaalude) and MDMA (Ecstasy). Such substances were allegedly produced for so-called 'crowd control', similar to how tear gas is used. At that time, there were significant unrest and large demonstrations from the black part of the population. However, there is reason to believe that these drugs were meant to be spread as narcotics in black neighbourhoods to destabilise the population. Additionally, the production of narcotics had a commercial purpose of bolstering the finances of Project Coast – and possibly also the private accounts of Basson and his colleagues.[56, 57]

A characteristic feature of the South African bioweapons programme is that about one-fifth of the research projects were aimed at means to reduce fertility in the black majority population. Basson himself was very concerned about this. A central project was the development of a fertility-inhibiting vaccine intended to be administered to black women. Some within the project hoped – albeit on highly uncertain scientific grounds – that such a vaccine could be made ethnically specific for blacks, however this research yielded no concrete results.

There are also indications that Basson encouraged his researchers to develop a so-called 'black bomb' – an ethnic weapon against blacks. Hence, Project Coast also had clear racial hygiene aspects. It has indeed been claimed that one could speak of genocidal traits in this programme.[58, 59]

South Africa would likely not have been capable of building its extensive programme for biological and chemical weapons without international assistance, regardless of Wouter Basson's energetic efforts. There has not emerged any evidence that any states officially supported the establishment of Project Coast. During his many foreign trips while leading Project Coast, Basson visited Europe, the USA, and Taiwan. There, he utilised his social skills to establish numerous contacts, including with researchers he met at conferences and with representatives of intelligence services. In this way, he may have acquired valuable information and assistance for the development of Project Coast.[56, 57]

Project Coast Disintegrates

In 1989, Frederik Willem de Klerk became president and leader of the Afrikaner-dominated National Party after Pieter Botha. De Klerk realised that the white, minority-dominated apartheid system was collapsing and needed to be dismantled. He did not have information about all of Project Coast's activities but had only been told that this programme focused exclusively on defence against biological and chemical weapons. Nevertheless, in March 1990, de Klerk ordered that all production of lethal biological and chemical weapons cease – but he still allowed work on means for 'crowd control', which he likely assumed was intended only for that purpose. At the turn of 1992–93, he then received a rather comprehensive and unvarnished report on Project Coast's dubious activities and was reportedly shocked. Shortly after this, Wouter Basson was quietly dismissed from his position, and it was decided that all of Project Coast's activities should be shut down.

Shortly after his dismissal, Basson was curiously reinstated in a position for 12 months to 'tie up the loose ends' of Project Coast. He was instructed to destroy all materials from the bioweapons programme. He may have destroyed the biological material – bacteria and viruses – however, he did not destroy the extensive documentation of the research activities and results. This was transferred to CDs that were stored secretly in a safe in a military building in Pretoria. What no one knew until several years later was that Basson had also made copies of all this valuable material for his personal archive.

At this time, the intelligence services MI6 in the UK and the CIA in the USA were well informed about the allegedly terminated bioweapons programme in South Africa and Basson's central role in it. Therefore, both Britain and the USA sent strong representations to de Klerk to ensure that all activities of the programme were discontinued. They also sought access to the documentation of Project Coast's activities.

After his final dismissal from the army, Basson had various assignments in the business sector. In this connection, he continued to undertake foreign trips. Among other things, he repeatedly visited Libya, Muammar Gaddafi's authoritarian, anti-Western regime. Both MI6 and the CIA increasingly monitored his travels because they feared he could sell his expertise to someone like the highly unpredictable Gaddafi, who was already pursuing chemical weapons. The USA and Britain now urged Nelson Mandela to offer Basson a position so that they could keep an eye on his activities. In 1995, Basson was appointed head of the cardiology department at a military hospital in Pretoria but he would not remain out of the spotlight.

The Truth Emerges

In 1995, Mandela's government established a so-called Truth and Reconciliation Commission (TRC) following the fall of the apartheid regime. By allowing both victims and perpetrators from the former regime and liberation movements to testify, the commission sought to shed light on the many violent acts and human rights violations that occurred during apartheid. The aim was not to convict individuals but to invite perpetrators who confessed and expressed remorse to seek amnesty. This was intended to heal the wounds.

In 1998, the TRC held a special hearing on South Africa's bioweapons programme. Many participants in Project Coast testified openly, providing the public with insights into the many shadowy actions that had been committed for the first time. Only at the very end of these TRC hearings was Basson interrogated. His testimony was brief and evasive. He did not admit to any criminal activity and therefore did not apply for amnesty.[56, 57]

In 1998, a comprehensive report from the TRC proceedings was published, in which a separate chapter provided a fairly detailed and unvarnished description of Project Coast's activities over many years. Many now hoped that the hour of justice had come for the man who had been dubbed 'Dr Death' in public. In 1997, he had indeed been arrested on suspicion of selling Ecstasy tablets, which had been produced in large quantities in Project Coast, as we have seen. However, the prosecution was unable to prove this. Instead, a remarkable discovery was made during a house search at Basson's residence: three suitcases containing documentation of the entire spectrum of activities over many years in Project Coast. Basson likely secured these documents for several reasons. He wanted to demonstrate that, as the leader of Project Coast, he had acted under orders from top military leaders. Furthermore, the documents might also serve as leverage against certain individuals. Finally, he may have been proud of the research activities he had led.[25]

This shocking discovery led to a trial against Basson that began in October 1999 in Pretoria. He was charged with no fewer than 67 criminal offenses, including numerous murders and murder attempts, fraud, embezzlement, and drug trafficking. The prosecution had 153 witnesses with damning testimonies against Basson.[56, 57]

On 22 April 2002, the court delivered its judgment in the case against Basson. To the surprise of the world, he was acquitted of all charges. This has later drawn strong criticism against the judge in the case, who was evidently biased. Desmond Tutu stated that this was 'a sad day for the credibility of the South African judicial system'. The verdict was appealed to the Supreme Court of South Africa, which partially accepted the appeal and opened the door for a new trial. However, this never happened.

In 2013, the Health Professions Council of South Africa concluded after seven years of investigation that Basson would lose his medical licence due to unethical behaviour but this also had no consequences for him. Basson disappeared from the spotlight until media in 2021 discovered that he was practising as a heart specialist at an upscale private clinic outside Cape Town.[60]

Basson fits into the category of doctors and researchers who have utilised their professional background to develop bioweapons without regard for fundamental ethical principles. However, he stands out in this group by actively contributing to the use of lethal biological weapons. Thus, it is not undeserved that he has been referred to in public as Dr Death and 'South Africa's Josef Mengele', a reference to the infamous Auschwitz doctor.

Biological Weapons in the Holy Land

On 29 November 1947, the UN decided that Palestine – then a British mandate – would be divided into one Jewish and one Arab state. The Arab states refused to accept this. After David Ben-Gurion declared the state of Israel established on 14 May 1948, five Arab armies attacked the newborn state. The Israeli army was poorly equipped but fought with desperate courage and managed to push back the Arab forces. In this war, which ended with a ceasefire in 1949, there is some evidence that Israel used biological weapons.[40, 61]

In his diaries and letters, Ben-Gurion – Israel's first Prime Minister – expressed his conviction that science and technology would be central to his new country's survival. Israel found itself in a difficult security situation with hostile neighbours on all sides.

As a student in Istanbul during World War I, Ben-Gurion witnessed the genocide of the Armenians. This experience, combined with the Nazi genocide against the Jews, undoubtedly shaped his view on the necessity of using all means – including biological weapons – in the Jewish fight for defence. A frequently cited statement from one of his letters is the desire to recruit Eastern European scientists who 'could either kill masses of people or heal masses of people; both are important'. This has been interpreted as an endorsement of the use of biological weapons in Israel's extremely challenging situation.[61]

Ben-Gurion had several competent scientific advisors who recommended the establishment of a dedicated research unit within the Israeli army. This 'corps of researchers' was designated as HEMED, an acronym for its Hebrew name. A strictly separate sub-division that was to work on biological weapons was subsequently attached to this unit. This special unit was designated HEMED BEIT. It was likely HEMED BEIT that was responsible for the use of biological weapons during the war from 1948 to 1949.

The establishment of HEMED BEIT and Israel's use of biological weapons was highly controversial among Israeli leaders, including the military. Many believed that biological weapons were 'dirty'. However, in light of the genocide against the Jews, the founders of this unit were convinced that any means could be justified in Israel's struggle for survival.

Until today, HEMED BEIT's activities during the 1948–49 war have been shrouded in strict secrecy from the government's side. However, in recent years, a couple of publications have shed some light on the use of biological weapons during 1948, even though official Israeli documents on this matter are still classified.

Israel's bioweapons operations against the Palestinians were code-named 'Send Your Bread Over the Waters' – a reference to the Old Testament, Ecclesiastes, chapter 11, verse 1. A significant part of this activity aimed to contaminate wells in Palestinian villages that served as bases for Palestinian soldiers with typhoid and dysentery bacteria. This was intended to prevent soldiers from returning to the villages and to increase the evacuation of the Palestinian inhabitants. A particularly well-known operation took place near the city of Acre, where the Israelis contaminated the old Roman aqueduct that served as the main water source. This led to over 90 cases of typhoid fever, with some fatalities among the residents, and resulted in increased evacuations that facilitated the Israelis' conquest of the city. Some British soldiers stationed in the area were also affected by the epidemic.

The Israelis possibly also used biological weapons against military forces from the attacking Arab states on at least one occasion. In May 1948, two Israelis allegedly attempting to contaminate wells with typhoid and dysentery bacteria were caught in the Gaza area where Egyptian soldiers were stationed. Confessions were likely obtained under torture and the two Israelis were later executed by the Egyptians. The Israelis categorically denied any involvement in this operation.

It is also known that the Israelis mapped wells and water pipelines in enemy capitals such as Cairo and Beirut, likely targeting them for bioweapons attacks. However, there is no evidence that such attacks were carried out.

There are few reliable details about Israel's relationship with biological weapons after 1949. The principle has been that the Israelis neither confirm nor deny having or having had weapons of mass destruction, including biological weapons.

However, it is a fact that HEMED BEIT, which was established to work with biological weapons, formed the core of the new state research institute Israel Institute of Biological Research (IIBR), which was created where HEMED BEIT had its laboratory in Ness Ziona, a few miles south-east of Tel Aviv. IIBR, which has grown significantly over the years, was initially granted both civilian research tasks and responsibility for research into biological and chemical weapons, which at the time was fully permitted under international agreements. IIBR, located in an orange grove, is protected by extensive security measures and doesn't even appear on maps of the area.[61]

There is no doubt that Israel has an advanced defensive programme against biological weapons. Whether the country has also developed biological weapons for military use, however, remains unclear and can only be the subject of speculation.

It is a fact that Israeli research in biotechnology and molecular biology is of international top quality. Publications from IIBR over the years show that much of the research is relevant to bioweapons initiatives. It has particularly focused on the bacteria that cause plague and typhus, on the rabies virus, and on various toxins. Although this itself is not proof, the general belief is that Israel possesses biological weapons, but probably not in large quantities ready for use.

As mentioned, all activities related to biological weapons in Israel and those at IIBR in Ness Ziona are strictly classified. It is interesting that one of the key scientists there – Avraham Marcus Klingberg – who was recruited in 1948 and was one of Israel's top experts on biological and chemical weapons, was arrested and convicted of espionage for the benefit of the Soviet Union in 1983. This was kept strictly secret for ten years, during which he served his long prison sentence under a false name, and his family was forbidden from informing others about his fate. Klingberg was almost a modern parallel to the legendary, nameless 'Man in the Iron Mask' – a person who, according to Alexandre Dumas, was secretly imprisoned by Louis XV and had to wear a mask to conceal his identity until his death. One might wonder what secrets Klingberg had revealed to the Soviet Union that made such extreme secrecy necessary. Could information from Klingberg have been useful for the Soviet bioweapons programme?[40]

It is perhaps not without reason that Israel – as one of very few states – has never signed the BWC.

Biological Weapons in Iraq – Lies and More Lies

Iraq emerged as an independent state in 1932. In 1979, Saddam Hussein, who belonged to the ruling Baath Party, came to power and quickly developed into a dictator. With support from both the Soviet Union and Western countries, Hussein built a military force that represented a significant threat in the region. In addition to conventional weapons, he focused on biological and chemical weapons and also wanted to develop nuclear weapons. From 1980 to 1988, Iraq was engaged in a devastating war with Iran – the First Gulf War. In this war, Saddam Hussein used chemical weapons. He also employed such weapons internally against Kurdish insurgents.

The Second Gulf War and Bioweapons

In 1990, Iraq invaded Kuwait. This led to the Second Gulf War – Operation Desert Storm – in which a coalition of 34 countries, backed by a UN mandate, liberated Kuwait.

Intelligence reports before the war indicated that Saddam Hussein had a comprehensive bioweapons programme. Therefore, coalition forces prepared for potential biological weapons attacks and were trained in the use of protective equipment, 150,000 personnel were vaccinated against anthrax, and 8,000 received a newly developed vaccine against botulin toxin. In addition, there was a significant stockpile of antibiotics at the front to be used in the event of an anthrax attack. Biological weapons were not used in the war, but veterans from this conflict later reported that the threat of bioweapons attacks from Iraq was one of the most psychologically burdensome aspects of their combat experience.[25, 34, 40]

There was also fear in Israel of attacks with biological and chemical weapons from Iraq during this war. The entire population was provided with gas masks and instructed to ensure access to sealed shelters with necessary supplies of food and water. During an Iraqi attack with Soviet-made Scud missiles, births occurred in Israel with gas masks worn by all involved, including both mothers and newborns.[40]

After the war, the UN Security Council mandated Iraq to ratify both the Geneva Protocol from 1925 and the BWC from 1972, as well as to destroy all biological and chemical weapons and cease all research and production activities in this field. They were also to destroy all rockets and missiles with a range over 150 kilometres. This was to take place under international supervision. The task was assigned to a UN commission – UNSCOM – led by the Swede Rolf Ekeus.[25, 34, 40]

This commission worked diligently in Iraq from 1992 to 1995 to find evidence of bioweapons. Participants received very little assistance from Iraqis, who for a long time denied that they had possessed such weapons. Detecting bioweapons was nearly like searching for a needle in a haystack, as facilities for the production of such weapons can be very difficult to locate without some clues as to their whereabouts. Nevertheless, Ekeus and his team gradually managed through meticulous detective work to find concrete indications that Iraq had indeed developed bioweapons. Confronted with this, the Iraqis initially admitted that they had had a rather modest bioweapons programme, primarily focused on defence against such weapons. Over time, however, they had to concede under the pressure of UNSCOM's new findings that Iraq had also produced various microbes for use as bioweapons, but that such weapons had never been completed.

Then UNSCOM achieved a breakthrough. In July 1995, General Hussein Kamal, Saddam Hussein's son-in-law, fled Iraq to Jordan, where he sought asylum. There he provided Rolf Ekeus with new information about Iraq's bioweapons. Iraq initially attempted to lure Kamal back, but when that failed, they made an unsuccessful attempt to assassinate him.

To pre-empt new admissions from Kamal, the Iraqis now acknowledged that the country had not only developed microbes for use in weapons but had also completed several types of bioweapons. Hidden in a chicken coop on General Kamal's farm, Ekeus found abundant documentation of an extensive bioweapons programme that had been active.[40]

It turned out that the Iraqis had conducted research on both anthrax and botulin toxin, of which they had significant stockpiles. They had also worked with certain viruses, in addition to fungal toxins, including a toxin that attacks the wheat plant. Strangely, they also focused on aflatoxin – a toxin that causes liver cancer, but only after a long time.[34, 40]

Furthermore, Iraq possessed bombs, missiles, and aerosol spray tanks for use on aircraft, all capable of disseminating anthrax, botulin toxin, and aflatoxin.

The Iraqis claimed that all bioweapons had been destroyed and that all activities in this field had ceased in 1991.

UNSCOM was also to prevent Iraq from resuming production of biological and chemical agents through repeated inspections. However, they were continually obstructed by the Iraqis, who also accused the commission of espionage against Iraqi military institutions. The extensive international sanctions against the country did not improve the working climate either. Nevertheless, the inspectors continued their work until they were evacuated at the end of 1998. UNSCOM concluded at that time that it was highly questionable whether Iraq had indeed terminated its bioweapons programme in 1991.[34]

A few days after UNSCOM was evacuated, President Bill Clinton ordered a four-day attack – Operation Desert Fox – which involved bombing Iraqi military institutions that could have been engaged in biological and chemical weapons.

The UN Security Council then established a new commission – UNMOVIC – to continue UNSCOM's work in Iraq. The Iraqis initially refused to cooperate, but after pressure from the UN, UNMOVIC, led by the Swede Hans Blix, was granted access to the country at the end of 2002. In the roughly four months that the commission was allowed to operate in Iraq, it conducted 350 unannounced inspections of suspected locations – but found no signs of biological and chemical weapons. Instead, they found and destroyed several rockets whose range was in violation of previous agreements.[34]

The Third Gulf War and the Propaganda of Lies

In March 2003, American and British forces attacked Iraq and started the Third Gulf War. This war was extremely controversial and did not have widespread support among European leaders. The new American president, George W. Bush, provided a two-part justification for the attack on Iraq: the country supported Islamist terrorist groups like Al-Qaeda, and Saddam Hussein had biological and chemical weapons. Therefore, Hussein's regime had to be removed. These claims were highly questionable.

On 5 February 2003, US Secretary of State Colin Powell spoke to the UN Security Council. He claimed that the US had evidence that Iraq possessed 'weapons of mass destruction', including biological weapons. As 'evidence', he held up a test tube that supposedly contained anthrax bacteria. He also waved aerial photographs that he alleged showed mobile bioweapons laboratories. Powell's assertions in this later infamous speech were unfounded, as shown by UNMOVIC's report. The speech had been prepared by the warmongering vice president Dick Cheney's office, where reports from the CIA had been processed. The so-called mobile bioweapons laboratories were in reality ordinary trucks.[34]

Powell would later bitterly regret this speech, describing it as 'a blot on my career'.[63] But it made a strong impression on the world.

Britain, under Prime Minister Tony Blair, wholeheartedly supported US attack plans and contributed military forces during the invasion of Iraq. The Blair government's follow-up on Powell's incorrect claims about bioweapons in Iraq created a serious crisis for the Prime Minister and his administration

– and had tragic consequences for one of the world's leading bioweapons experts, David Kelly.[34, 64]

The Kelly Affair

To defend the attack on Iraq, the British government issued a statement in May 2002 regarding the Iraqi threat. It stated that Iraq could attack with biological or chemical weapons 'only 45 minutes after an order had been given'. A reporter at the BBC claimed in May 2003, based on an interview with an unidentified bioweapons expert, that the Blair government had dramatised ('sexed up') uncertain intelligence data and essentially lied to the British public to justify Britain's participation in the Iraq war.

A fierce conflict arose between the BBC and the government. The BBC indicated that their source was David Kelly. Kelly was Britain's leading bioweapons expert and had worked for many years at Porton Down, the British centre for bioweapons research. Additionally, as a member of both UNSCOM and UNMOVIC, he had extensive experience from Iraq. He had also actively participated as an inspector in the dismantling of Biopreparat after the fall of the Soviet Union.

It is likely that the BBC reporter had substantially overstated what Kelly had told. However, Kelly found himself in the crossfire. He faced severe criticism from both his superiors in the Ministry of Defence and from a parliamentary committee for allegedly inappropriate and disloyal behaviour. In the heat of the moment, he also gave some incorrect information about his contact with the press. All of this deeply affected him. It ended with Kelly being found dead with a severed artery on his left wrist at his favourite spot on a small clearing in the woods near his home in idyllic Oxfordshire.[65]

Both forensic investigations and a special inquiry commission concluded that Kelly had committed suicide. His alleged motive was despair over the accusations that led to loss of esteem and possibly even the loss of his pension.[64] However, there were a number of striking findings regarding his death that have led some to claim that it could not have been suicide, but murder.[65] The motive was said to be that Kelly held secrets from his many years of work on bioweapons that he intended to reveal in a book he was rumoured to be writing. According to proponents of the murder theory, some sought to prevent this. Shortly before his death, Kelly sent an email to the

American journalist Judith Miller, who was working on bioweapons,[52] in which he wrote that 'dark actors' were after him.

The Kelly affair resulted in a loss of reputation for many of those involved and led to dismissals at the BBC. There is also no doubt that it significantly damaged Tony Blair's reputation. At press conferences, he was asked whether he 'had blood on his hands'.[64]

In retrospect, it is beyond doubt that the Blair government's statement, which triggered the Kelly tragedy, contained clearly erroneous claims about Iraq's biological and chemical weapons. Whether the responsibility for this rests with the government or the British intelligence service remains unclear.

Chapter 7

After the Fall of the Wall – Bioweapons After the Cold War

After the anthrax disaster in Sverdlovsk in 1979, Western countries had strong suspicions: the Soviet Union had a bioweapons programme developing weapons that were clearly in violation of the BWC (Biological and Toxin Weapons Convention) that the Soviets had signed. As the 1980s progressed, intelligence agencies gathered more and more evidence supporting this suspicion.

Under the last president of the Soviet Union, Mikhail Gorbachev, a liberalisation process began, with key words being *perestroika* (restructuring) and *glasnost* (openness). From 1990, considerable political pressure was exerted by the USA and the UK on the Russians to obtain clear admissions of their extensive bioweapons programme.[34]

At the end of 1989, a key figure in the Russian bioweapons programme, Vladimir Pasechnik, fled to the UK, providing important information about the bioweapons situation in the Soviet Union. He had led one of the research institutes in Biopreparat and could provide detailed information about the organisation and many of the research projects. This was new and shocking information for the Western powers. A key person during the interviews with Pasechnik was the previously mentioned David Kelly.[67]

President George Bush in the USA and Prime Minister Margaret Thatcher in the UK confronted Gorbachev with Pasechnik's information, which led to Gorbachev inviting British and American representatives to inspect several facilities within Biopreparat. Some inspections were conducted, but they were generally unsatisfactory. Inspectors often received misleading and outright false information, and on some occasions they were denied access to parts of the facilities that were clearly central to bioweapons production. Therefore, it was necessary to establish clear and formalised agreements for inspections. This led to a tripartite agreement in 1992 between the newly formed state of Russia on one side and the USA and the UK on the other.[68]

In the spring of 1992, another researcher from Biopreparat defected to the West. His identity has been kept secret, and he is referred to only by the code name Temple Fortune. He worked at a lower level in Pasechnik's institute but could confirm his information. Additionally, he could provide information on developments in the period following Pasechnik's flight.[67]

By the autumn of 1992, the USA and the UK had even better leverage in negotiations with Russia. In that year, Kanatzhan Alibekov fled to the US from Russia.[34, 69] He was a microbiologist and had been second in command throughout the Biopreparat system. He later took the more English-sounding name Ken Alibek. He could offer even more extensive information about the Soviet Union's bioweapons programme than Pasechnik. He described the entire Biopreparat organisation – facilities, personnel, research activities, and practical solutions for using the developed bioweapons. This was also detailed in his 1999 book *Biohazard*.[33]

According to the tripartite agreement between Russia, the UK, and the USA, mutual inspections of bioweapons facilities were to take place, but Russia chose to exclude all military institutions. This was already a significant weakness, as military facilities had been an important part of Russian bioweapons efforts. The agreement also contained a number of other provisions regarding cooperation on bio-defence and the exchange of researchers and information in the field. However, this was never realised.

From 1994, several inspections were carried out pursuant to the aforementioned tripartite agreement. Visits in Russia showed clear signs at several of the visited facilities that bioweapons development had occurred there. The Russians were also shown several centres in the USA that had been engaged in bioweapons work. They also expressed interest in seeing some facilities from the pharmaceutical company Pfizer and received permission for this from the American government. However, this led to strong reactions from Pfizer and the American pharmaceutical industry in general. They argued that this was a threat to commercial secrets. The Russians reacted to this, believing it indicated secret development of bioweapons.

Gradually, the willingness to cooperate on which the agreement was entirely dependent deteriorated. The openness of the Russians regarding both past and current bioweapons activities proved to be limited. From 1996 onwards, the tripartite agreement was effectively dead.[68]

President Boris Yeltsin admitted in 1992 that the Soviet Union had had an offensive bioweapons programme. He also declared that he had ordered the termination of this programme. It is highly doubtful that this occurred to the full extent, although some bioweapons facilities were destroyed while the cooperative agreement was in effect.

Western inspectors gradually gained access to facilities within the Biopreparat system and met scientists who had been involved in the bioweapons programme. They were told that salaries were no longer being paid, which led to significant economic problems for the scientists who now found themselves without jobs. At the same time, they reported that shortly after the dissolution of the Soviet Union, agents from Iran and North Korea had appeared with offers of well-paid positions in their respective countries. Thus, a serious situation arose where many Russian bioweapons researchers were likely to emigrate and take with them both their knowledge and materials that had been developed as part of bioweapons development.[34]

This serious problem was sought to be addressed through the so-called Cooperative Threat Reduction Programme. This was based on a law that the two American senators Sam Nunn and Richard Lugar had passed in the American Congress in 1991. The programme, which received significant funding from the USA, was intended to ensure the safe destruction and eventual storage of nuclear weapons in Russia and other former Soviet republics. However, the programme was quickly expanded to also include biological and chemical weapons.[68]

Through this programme, which is considered successful, changes were made to the tasks at research institutions in the former Biopreparat network in a civilian direction, and retraining and new positions were secured for a large number of the former bioweapons scientists. In this way, a significant 'brain drain' that could have occurred to Iran and other states wishing to establish their own bioweapons programmes was largely prevented. The authorities in the new Russia cooperated for several years with the Americans on this programme, which had its own office in Moscow. But in 2014, the cooperation ceased, and the office was moved to Kazakhstan. Thus, the outside world lost its last opportunity for some insight into Russian research with potential relevance for bioweapons development.

Bioweapons in Russia Under Vladimir Putin?

Regardless of previous admissions, the official Russian stance since the late 1990s has been that the Soviet Union never developed bioweapons but only worked on bio-defence.

As mentioned, Western inspectors were never granted access to the military research institutions involved in the Soviet Union's former bioweapons programme. What actually happens at these institutions has been shrouded in total secrecy in the years following the dissolution of the union. What we know is that, at least at the three most important of these institutions, significant resources are being invested in modernisation and new construction. Furthermore, much suggests that advanced biotechnological research is being conducted there. In 2006, the various military institutions that had worked previously with biological weapons were merged into one organisation – the 48th Central Research Institute (48. CSRI). This organisation collaborates with the intelligence agency FSB on, among other things, chemical weapons.[68, 70]

Officially, only research on defence against bioweapons is being conducted in Russia today. This includes the development of vaccines, where military laboratories are clearly cooperating with civilian institutions. This includes vaccines against the Covid-19 virus, with Russia being the first country to approve such a vaccine.

But is Russia also focusing on the development of biological weapons as part of its military arsenal? This cannot be stated with certainty, but many suspect so. As we have seen, the Soviet Union secretly built up its bioweapons programme in direct violation of the BWC. On several occasions, American authorities have officially claimed that they believe Russia has such a programme. Over the years, there have been several hints from the Russian press, and on certain occasions, from central Russian officials about activities in the field of bioweapons.

In 2012, Vladimir Putin gave a speech after being elected president for a second time, where he spoke about new forms of weapons. He mentioned 'genetic weapons' that Russia wished to develop. He was likely referring to bioweapons. Following Putin's speech, Defence Minister Anatoly Serdyukov stated that he would follow up with 'the development of new weapons based on physical principles', including 'genetic weapons'.[47]

A notable aspect of the Russian discussion on 'genetic weapons' is the interest in 'ethnic weapons' – bioweapons that allegedly selectively target certain ethnic groups based on their genetic profile. It is worth noting that Putin, in 2017, banned the export of biological materials from ethnic groups in Russia. Most scientists believe that such weapons are currently unrealistic. Nevertheless, Putin's stance indicates a focus on biological weapons.[68]

We still lack documentation proving that Russia has an active bioweapons programme. However, it is a fact that the country is investing significantly in biotechnological research. In certain areas, Russian scientists have undoubtedly made considerable progress, although this research has likely not reached the same level as in, for example, the USA and China.

It is possible that the war in Ukraine and the extensive Western sanctions against the country have led to resource problems for the biotechnological research efforts in Russia. If Russia indeed has an active bioweapons programme, one might think that this would be prioritised if such weapons are deemed useful for the country's military effort, as conventional weapons have so far not led to any decisive success in the war in Ukraine.

The USA and Bioweapons After Nixon's Decision in 1969

President Nixon decided in 1969 to terminate all activities aimed at developing and storing bioweapons and to focus solely on defence against them. However, toxins were not included. After input from scientists, Nixon chose to include such biological weapons a few months later.[34]

The background for Nixon's decision, which came as a surprise to those who knew that the USA now had a significant and proven bioweapons arsenal, was probably, as mentioned previously, a broad political reaction with increasing opposition to bioweapons. At the same time, Nixon had been convinced by several prominent bioweapons opponents that such weapons would not be of significant use now that the USA had an extensive arsenal of nuclear weapons. At one point, he stated: 'We'll never use those damn germs – If someone uses germs on us, we'll nuke 'em.'

In Nixon's statement, it was also said that 'biological weapons have massive, unpredictable, and potentially uncontrollable consequences. They can trigger global epidemics and harm the health of future generations. Humanity already holds far too many seeds for its own destruction. By the examples we provide today, we hope to contribute to an atmosphere of peace and understanding among all nations.' This suggests that there was hope that the USA's decision would inspire other states to eliminate any bioweapons programmes.[34]

Bioweapons scientists in the laboratories at Fort Detrick were likely seething with anger over Nixon's statements, which they perceived as purely political. The fact was that through extensive research, they had developed microbes and methods of dissemination that were not 'unpredictable' and 'uncontrollable', as stated in Nixon's remarks. The microbes that had finally been used in completed weapons would not 'trigger global epidemics', as many either had very low infectivity from person to person or, like anthrax, were not transmitted between humans at all. The idea that future generations would pay the price was also regarded by researchers as a fabrication.

But Nixon's decision had to be followed. Several scientists in the bioweapons programme lost their positions. Parts of Fort Detrick that had been important for bioweapons testing were destroyed. Extensive work was initiated to destroy safely the large stockpiles of microbes and finished biological weapons that had been stored at the facilities. The result of over 25 years of research was now destroyed, and the facilities also had to be disinfected. Overall, this was a substantial effort that took a couple of years.[34] But was everything of microbes and toxins destroyed in this process?

In most countries that have so far invested in the development of bioweapons, there has been a close collaboration between engaged research environments and intelligence agencies. This applied significantly to the USA as well. The CIA had major interests in this field, as they needed such weapons for part of their operations, which have always remained highly classified. This includes secret political assassinations, where the CIA has a murky history.

The CIA already had its own stockpiles of microbes and toxins for weapon purposes. They did not want to destroy them following Nixon's decision, believing it could not apply to intelligence services. It later became clear that while most of the CIA's stockpiles were indeed destroyed, they maintained a stock of saxitoxin – derived from shellfish – sufficient to kill 55,000 people.[34]

From then on, the official stance was that the USA's bioweapon efforts would focus solely on developing an effective defence against such weapons. This has undoubtedly been important. But if President Nixon had hoped that other countries would follow the USA's lead and refrain from investing in bioweapons with his decision in 1969, that was unrealistic. Even after the BWC was adopted in 1972 and came into force in 1975, it has turned out that interest in bioweapons has hardly diminished – not even in all the countries that signed the agreement.

Just as the USA's active bioweapons programme before 1969 was largely kept secret, there has also not been full transparency about the country's biodefence programme since. Only a limited group of scientists at Fort Detrick, with the CIA as a significant collaborator, as well as a few congressional members and key bureaucrats in the US security agencies, have had insight into this work.

In the years following 1975, there was little enthusiasm among American politicians and leading military officials for greater investments in work against bioweapons. Information about other countries' investments in the field was consistently rather vague, although the previously mentioned Sverdlovsk incident raised some suspicion that the Soviet Union had biological weapons. However, initially, as we will return to, leading American bioweapons experts like Matthew Meselson chose to interpret the anthrax outbreak in Sverdlovsk as a natural epidemic. In retrospect, it is likely that Meselson – one of the proponents of Nixon's decision in 1969 and for the USA's engagement in the BWC – struggled to acknowledge that this international agreement was not effective.[28]

After a congressional investigation in 1990 regarding the ongoing bioweapons programme, the conclusion was that the military scientists at Fort Detrick had used too many resources on unnecessary research. Additionally, it was believed that research in certain areas had approached the limits of what was permitted under the BWC. In the following years, bioweapons scientists at Fort Detrick largely adhered to 'approved' projects.

Increased Fear of Bioweapons

When revelations about Iraq's extensive bioweapons programme emerged in 1995, the tide began to turn. The increasing frequency of terrorist attacks, including from fanatical Islamists like Al-Qaeda, undoubtedly contributed to increased attention to the threat from bioweapons during President Bill Clinton's administration.[34]

Clinton himself became deeply concerned about bioweapons. He had several prominent scientists as advisors, including Nobel Prize winner Joshua Lederberg and Craig Venter, who was central to mapping the human genome. Both strongly urged him to invest in biotechnology research for defence against biological weapons. Interestingly, it was likely a thriller – Richard Preston's *The Cobra Event* – that was crucial for Clinton's engagement.[28, 52] Preston had attracted attention previously with his book *The Hot Zone* about the terrifying properties of the Ebola virus. He had secured background information from various researchers and military personnel knowledgeable about bioweapons. *The Cobra Event* describes how a deranged genius triggers a bioattack in New York using a 'designer virus' – a combination of smallpox virus and two others. The result is a deadly, highly contagious microbe that leads to a horrific infection where the brain disintegrates rapidly.[71]

The result of this development was that over the next few years, there was an escalation of the bio-defence programme. Increasingly, the Americans sought to study not only relevant microbes but also finished bioweapons, primarily bombs. Previous information from defectors like Pasechnik and Alibek had shown that the Soviet Union had made considerable progress in developing such bombs. The CIA, which was heavily involved in the field, attempted to obtain a copy of the Soviet bomb and succeeded, at least in finding parts of one. Researchers at both Fort Detrick and the CIA believed that they needed to build similar bombs to test their effects in order to plan defences against them. This work was named Project Clear Vision.[52]

With this project, they had again approached dangerously close to the limit of what the BWC permits. The CIA's lawyers believed it was clearly within the boundary, but not everyone familiar with the project agreed.

This discussion persisted. Some experiments were conducted with bombs, but in 2001, Project Clear Vision was terminated, primarily due to lack of funding.

During these years, however, the CIA had several projects that challenged the boundary between defensive and offensive bioweapons research. In the strictly secret Project Bacchus, they sought to see if it was possible to build a laboratory to produce microbes for weapons using only commercially available materials. Again, the focus was on anthrax, although harmless, related bacteria were used. Within a year, they successfully built a laboratory for mass production of anthrax. This experiment demonstrated that states or terrorists with relatively limited resources could acquire bioweapons.

In Project Jefferson, they aimed to develop a 'superbug' – vaccine-resistant anthrax bacteria – through genetic manipulation, as Soviet researchers allegedly had done after the Americans attempted unsuccessfully to secure this bacterium. It is unclear what the outcome of this project was, as it remained strictly confidential due to fears of reactions.

According to the BWC, the USA and other states that signed the agreement were obligated to give an annual report on activities in their bio-defence. Interestingly, none of the aforementioned projects were reported.

What Does the USA Have to Hide?

In 2001, a number of the states that had signed the BWC were inclined to strengthen the control provisions in the agreement. This particularly concerned increased opportunities for inspection of individual states' facilities and activities related to bio-defence. It came as a shock to many that the USA resolutely refused to accept this proposal.

The justification for the USA's dramatic decision did not seem convincing. Three reasons were given. Firstly, they argued that the inspection proposals would not be effective, even though they would clearly lead to increased transparency in bioweapons matters. Furthermore, the USA argued that increased transparency would undermine the necessary protection of industrial secrets, particularly in the pharmaceutical industry. This problem could undoubtedly have been resolved. Finally, the Americans claimed that the proposals for increased inspections would endanger the USA's bio-defence programme. This was also a surprising objection. What did they really have to hide?

Thus, questions arose about what was so threatening in the proposal to strengthen the BWC for the USA. Some of the answers likely emerged when three 'investigative journalists' in mid-2001 released the book *Germs – The Ultimate Weapon*, which revealed the three CIA-inspired projects mentioned earlier.[52] This showed that the USA granted itself freedoms that are not publicly acceptable in the interpretation of international agreements like the BWC – freedoms that would likely not have been allowed for other states.[34,52]

Suspicion arose – which still persists – that the USA may not only be engaged in bio-defence but also in the development of bioweapons. It is known that there has been a military desire to amend the BWC so that it could permit the development of bioweapons using genetically modified microbes that could target non-living military objectives such as coverings, camouflage paint, fuel, and similar. Furthermore, there has also been a desire to develop non-lethal biological weapons, while the language in the BWC undoubtedly prohibits this.[72]

On 11 September 2001, New York was struck by terror. A few weeks later, there was another terrorist attack – this time with anthrax. This led to a significant increase in resources for bio-defence, and the institutions central to defending against terrorist attacks were reorganised. A large portion of this funding has gone to basic biotechnological research. The National Institute for Allergy and Infectious Diseases (NAID) in Maryland, which for many years has been led by Anthony Fauci, played a central role in the American bio-defence programme. At the same time, both private pharmaceutical industry and several academic research institutions have been linked to the programme through contracts with the government.[34]

In 2018, the American Department of Defense launched plans for a new project – Insect Allies – aimed at protecting the USA's grain production from plant diseases and bioweapons attacks. Insects were to be used as carriers of genetically modified viruses that would quickly change the chromosomes of the plants. The project was criticised by several researchers. They claimed that it not only posed significant ecological risks but could also be used for the development of biological weapons.[73]

Accusations against the USA of bioweapon development in violation of the BWC have recently flared up in connection with the war in Ukraine. It is primarily Russia and China that have renewed accusations that have been raised repeatedly in the UN. Again, reference is made to the USA's persistent resistance to fortifying the BWC with increased controls. We will return to this conflict.

Chapter 8

The Shadow Play on Bioweapons – Which States Have Bioweapons Today?

State bioweapons programmes have always been shrouded in strict secrecy. This applies to democratic as well as totalitarian and authoritarian states. As a result, it has been challenging for the outside world to gain insight into any development of biological weapons. Historically, this uncertainty has actually contributed to the development of such weapons, as states have wanted to respond to potential bioweapons attacks with their own. We saw this during World War II and in connection with the Soviet Union's post-war build-up of its bioweapons programme, as Soviet leaders assumed that the USA had such weapons. The lack of insight into the bioweapons situation in various countries remains very much the same today.

Despite this, the public and politicians in most countries have, at least officially, shown little interest in the possibility of the use of bioweapons in military conflicts. This also includes the threat from bioterrorism, i.e., the use of bioweapons in terrorist attacks, although the anthrax attack in New York in 2001 briefly brought biological terrorism to the forefront.

This lack of interest in the threat from bioweapons is not only incomprehensible – it is also dangerous. Today, we must acknowledge that several states undoubtedly have bioweapons programmes of varying scope, but the official information on this is extremely sparse. The intelligence services of individual countries likely possess significantly more information that is not disclosed.

Over the years, American authorities have regularly issued reports on the global situation regarding biological weapons, listing the states that are most likely investing in such weapons. It is mostly the same countries that keep coming up: Russia, Israel, Iran, North Korea, China, and Syria, as well as previously Iraq. The USA itself is, of course, not included on the list.

We have already described the situation in Israel and the evidence for a secret bioweapons programme there. The description of the situation in Russia after the Cold War also hints that the Russians are still investing in bioweapons, but this is after all only circumstantial evidence. What is the situation in the other states that the USA continually designates as 'bioweapon offenders'? This includes states like North Korea, Iran, and Syria, which Americans like to call 'rogue states' – states that behave unpredictably and irresponsibly in an international context, making them dangerous to other countries. However, as mentioned previously, there are few facts about bioweapons available to the public.

North Korea – Perhaps Not Just Nuclear Weapons?

It is no secret that North Korea has developed nuclear weapons and is thus a nuclear power. Such weapons are displayed openly during military parades in Pyongyang. The country has also developed intercontinental missiles that could potentially carry nuclear warheads. The world is – rightly – concerned about this threat from North Korea. Due to its nuclear weapons programme, North Korea has been subjected to international sanctions.

However, there has been little discussion about the possibility that North Korea also possesses biological weapons, even though the country has long been on the list of suspected countries. Both the USA and South Korea, which understandably are particularly concerned about potential threats from North Korea due to geographical proximity, have over the years made somewhat contradictory statements about the country's possible bioweapons – partly categorical, partly more vague – stating that the North Koreans might have such weapons. In the USA, we have also seen diverging claims about this from different institutions.[74]

What is certain is that North Korea has increased its activity in modern biotechnology significantly in recent years. This could in itself pave the way for the development of biological weapons. Researchers from North Korea have also collaborated with Chinese and Russian researchers. South Korean intelligence has allegedly identified at least ten biotechnological laboratories where bioweapons may be under development. However, it is somewhat uncertain whether this is indeed happening.[75]

Several defectors from North Korea have asserted that the country has bioweapons – and they have even claimed these have been tested on humans. Such allegations should be evaluated critically, as defectors may have various motives to exaggerate such accusations and possibly demonise the leadership in North Korea.

Andrew C. Weber – one of the top bureaucrats in the USA who has led the country's defence programme against bioweapons and other weapons of mass destruction – stated in 2019: 'It is far more likely that North Korea will use biological weapons than nuclear ones.' He further asserted: 'The programme is advanced, underestimated, and very deadly.'[75]

Among the microbes that have been particularly feared are anthrax and smallpox viruses, which North Korea is believed to have stockpiled in violation of international agreements. It is unclear how such microbes would be disseminated in a military conflict. Some have mentioned the possibility that the country could use living carriers of deadly microbes such as the smallpox virus in a military conflict.

The Middle Kingdom and Bioweapons

No country in the world has suffered as much from bioweapons attacks as China. For many years before and during World War II, the Japanese army ravaged China, significantly aided by Ishii Shiro and his colleagues with their inhumane bioweapons programme. According to official Chinese sources, at least 270,000 military and civilian Chinese lost their lives due to Japanese bioweapons attacks. Additionally, China believed that the USA used bioweapons against them during the Korean War, a matter we will revisit. Therefore, it is not surprising that China later sought to build up a bio-defensive capability. This began in the 1950s, and the Chinese are open about it.

On the other hand, China has always denied that the country has an offensive bioweapons programme.[76] In 1952, China signed the Geneva Convention and in 1984 the BWC. Despite this, many have argued that China has had an active bioweapons programme since the 1980s. The USA has repeatedly claimed this officially and urged China to dismantle what they believe to be an offensive programme. However, it is somewhat unclear whether these repeated accusations are based on concrete intelligence or merely speculation.

The Soviet defector Ken Alibek has stated that the Russians were convinced that two epidemics of haemorrhagic virus near the lake Lop Nor in Xinjiang in the 1980s were due to leaks from laboratories that were working with viruses as weapons. Alibek's information has often proven to be accurate.

There is no doubt: China has made impressive advancements in a wide range of fields in research and technology in recent years, and development in fundamental and applied biotechnological research has been formidable. In these areas, China is today among the leaders in the world. Therefore, it is clear that the country possesses both the expertise and infrastructure needed to develop state-of-the-art bioweapons.

It is worth noting that Chinese scientists have invested heavily in research relevant to aerosols and airborne transmission, known as aerobiology – a field that is crucial for the effective dissemination of bioweapons.[77]

Those who believe that China has bioweapons also emphasize another factor, namely the close collaboration taking place between the military – the People's Liberation Army – and civilian research laboratories. The civilian laboratories have extensive contact with international research environments, and Chinese guest researchers are active in many parts of the world, not least in the research communities of the USA. In this way, data is obtained from cutting-edge research that could be very useful in a potential bioweapons programme.

The People's Liberation Army is also heavily engaged in biotechnological research and has several advanced research environments. Military publications from China often emphasize the importance of biology for future warfare. In a relatively recent military textbook, the possibility of new forms of biological warfare is discussed, including 'specific, ethnic genetic attacks'.[76, 77]

Everything is actually in place for an advanced bioweapons programme in China – including offensive intentions. With China's new role as a global superpower, it is clear that this possibility cannot be ignored. However, solid evidence for the existence of Chinese bioweapons is not available to the public.

In the tense situation between the People's Republic of China and Taiwan, biological weapons have also become a factor. In Taiwan, there are fears that

China may use bioweapons in an attack, and the military forces on the island conduct drills in anticipation of this.

Do the Ayatollahs in Iran Have Bioweapons?

In 1979, Muslim leaders led by Ayatollah Khomeini came to power through a revolution. The country has since created significant international problems. Iran is included on the unofficial list of 'rogue states' – countries that pursue unpredictable policies that the international community believes pose a threat to other countries. This includes not only Western countries but also Israel and several states in the Middle East that have Sunni Muslim regimes, in contrast to the Shia Muslim regime in Iran.

From 1980 to 1988, Iran and Iraq fought a bloody war in which both sides used chemical weapons. Iran has conducted nuclear research for many years. The general perception in Western countries has been that this research has not only been for civilian purposes but has aimed at the development of nuclear weapons. This has created significant fear, and attempts have been made to reach agreements with Iran to prevent this development. Various sanctions have also been attempted to pressure the country.

The possibility that Iran may also have biological weapons has, however, been overshadowed by the fear of nuclear weapons and efforts to halt Iran's development of these. But this could potentially have risky consequences. If efforts succeed in stopping Iran's nuclear weapons programme, the country may turn even more towards other unconventional weapons – such as biological ones.

Iran has ratified the BWC, nevertheless, Iranian leaders have expressed positive attitudes toward bioweapons. In 1988, Ali Rafsanjani, a leading politician and Ayatollah who later became president, stated: 'Chemical bombs and biological weapons are the poor man's atomic bombs and are easy to produce. We should at least consider them in our defence.' At that time, Iran had accused Iraq of using bioweapons in the protracted war between the two states, likely without reason.[78]

For many years, American authorities have included Iran on the list of states possessing bioweapons. However, similar to Iran, there are very few concrete facts in the public domain that support this assertion. Again, we must rely on circumstantial evidence and rumours.

It is a fact that Iranian agents were very active at the time of the Soviet Union's dissolution to secure researchers from the extensive Biopreparat programme. However, we do not know how many of these experts accepted such favourable offers from Iran.[78]

There is also no doubt that Iran has built up both expertise and biotechnological research environments over the years, which are necessary for the development of biological weapons. The country also has a well-developed pharmaceutical and biotechnological industry producing both vaccines and other biological products.[79]

The situation in Iran is thus the same as in several other countries where secret bioweapons development could be taking place: well-developed biotechnological research environments can support both legitimate research for civilian purposes and concealed bioweapons research. The lack of effective inspection mechanisms in international bioweapons agreements makes it practically impossible to reveal any dubious research.

Chapter 9

Dangerous Waters – International Agreements on Bioweapons

Biological weapons are classified within the group of weapons of mass destruction, along with nuclear, chemical, and radioactive weapons.[80] This group is often summarised as CRBN weapons (Chemical, Radiological, Biological, Nuclear). While nuclear and radioactive weapons are newcomers to this group, both biological and chemical weapons have a long history. Before the bacteriological revolution at the end of the nineteenth century, when the role of microbes as the cause of infectious diseases was discovered, there was not a clear distinction between biological and chemical weapons, although there was some awareness that certain diseases were contagious.

As we have seen, there are many historical examples of the use of bioweapons. Nevertheless, it is clear that the prevailing perception has been that such weapons are not 'civilised' and that their use is dishonourable. Biological weapons were only used by barbarians, such as the Scythians – not by 'civilised' peoples like the Greeks and Romans. The Roman author Valerius Maximus, who lived in the first century AD, succinctly and forcefully articulated this view: '*Armis bella non venenis geri*' – 'War is conducted with weapons, not with poison'. This attitude toward bioweapons still applies today. While all forms of weapons of mass destruction are regarded with fear and disgust, this is particularly true for biological weapons.

As early as the late nineteenth century, when the chemical industry was developing rapidly, people became aware of the potential danger from chemical weapons. This led to attempts at international agreements to prohibit such weapons, such as the Brussels Conference in 1874 and the first and second Hague Peace Conferences in 1899 and 1907. These rather simple agreements without monitoring mechanisms proved to be ineffective, for during World War I, poisonous gas was used by both sides, first at Ypres in 1915. It is estimated that 90,000 soldiers lost their lives due to chemical warfare, and many sustained permanent health injuries.[81]

When the international agreements on chemical weapons were established, there was no consideration given to including biological weapons. Recognition of the threat from such weapons came only some time into the twentieth century. In the first decades following the bacteriological revolution, the focus was solely on studying the microbial world to prevent and treat infectious diseases. However, as early as World War I, both Germany and France were engaged in developing biological weapons, and Germany deployed them.

The Geneva Protocol – Well-Intentioned but Ineffective

It was, however, the experiences with chemical warfare during World War I that led to the so-called Geneva Protocol in 1925, which initially concerned the prohibition of the use of such weapons.[34] After proposals from Poland, biological weapons also became part of the agreement, prohibiting 'the use of bacteriological methods in warfare'. The Geneva Protocol, which came into force in 1928, has been signed by a large number of states, although not all countries' national assemblies have ratified it.

The Geneva Protocol has played an important role by signalling that the international community does not accept the use of biological and chemical weapons. However, the agreement has a number of shortcomings that have significantly weakened its effectiveness.[34]

The protocol prohibits the use of the relevant weapons in war – but not research and development of biological weapons or the stockpiling of such weapons. It does not apply to states that have not signed it. The protocol only prohibits the use of the relevant weapons between states, not internally within individual states. Furthermore, many states that have signed the protocol have been granted a number of reservations. For instance, they reserve the right to respond with biological and chemical agents to attacks using such weapons. In practice, the Geneva Protocol thus only prohibits the first use of these weapons. Finally, this agreement has no provisions for monitoring or control of member states' behaviour regarding the use of biological weapons.

The many loopholes in the Geneva Protocol undoubtedly contributed to many states that had signed it actively building up bioweapons programmes in the following years – and in the case of Japan, significantly employing such weapons.

In the decades following World War II, as we have seen, there was an arms race in several states involving bioweapons development, especially in the USA, the Soviet Union, and the UK. However, in the second half of the 1960s, a shift in attitude toward biological weapons occurred. An important factor was undoubtedly the development of nuclear weapons. Many military and politicians believed that these potent weapons now rendered biological weapons redundant. Moreover, bioweapons had been somewhat untested in war and were perceived by many to be more unpredictable than both nuclear and conventional weapons.

This change in mood led the UK in 1968 to propose a new international agreement to prohibit bioweapons – an agreement that was intended not to replace but to supplement the Geneva Protocol. This proposal was approved by the UN General Assembly. Then came President Nixon's surprising decision in 1969 to dismantle the USA's bioweapons programme. This development led to the new international agreement that is commonly referred to as the Biological Weapons Convention (BWC), although its full name is the Biological and Toxin Weapons Convention (BTWC).

BWC – Important but Incomplete

The BWC was first signed by the USA, the Soviet Union, and the UK on 10 April 1972, in ceremonies held in London, Moscow, and Washington. Today, 185 states have both signed and ratified the convention. Four states have signed but not ratified it, while nine states, including Israel, have not signed the treaty.

The BWC was an important step forward in the efforts to prevent the use of bioweapons. In the preamble of the treaty, which has 15 articles, it is stated that the signing states are 'determined, for the sake of all humanity, to exclude completely the possibility that bacteriological (biological) agents and toxins will be used as weapons' and further states that they are convinced that 'such use would be morally repugnant to the conscience of mankind and that no effort should be spared to reduce this risk'. Here we see an expression of the moral condemnation that bioweapons have faced historically.[34]

Does the real content of the BWC align with the pathos found in the preamble? What does the treaty actually contain? From the outset, critics pointed out that it had serious shortcomings.

The core content of the BWC is the prohibition on the development, stockpiling, production, or transfer of biological agents and toxins 'of types and in quantities' that are not justified for protective or peaceful purposes. Furthermore, the treaty prohibits the development of weapons, equipment, and delivery systems to spread such agents and toxins.[34, 81, 82]

However, a major problem with the BWC is that it does not contain any provisions that allow for monitoring and effective control of member states' compliance with treaty obligations. Here, it follows in the footsteps of the Geneva Protocol. The only thing that has been agreed upon over time is that states shall voluntarily submit an annual report on relevant activities. Not all states do this, and the reports are often quite empty. Over the years, efforts have been made to address this significant weakness in the treaty.[34, 82]

The working method under the BWC is that representatives of the participating states meet every five years for 'Review Conferences' to discuss any amendments. After several years of work on an amendment proposal, which was then presented at such a conference in 2001, this was definitively halted by the USA, as we have seen previously. The USA has also subsequently, up until the last Review Conference held in December 2022, opposed such changes to the BWC that would increase the possibilities for monitoring and controlling the provisions of the treaty.[34, 82]

If states within the BWC group suspect breaches of treaty provisions, they can today only raise the issue with each other or refer it to the UN Security Council. The Security Council can then follow up with sanctions against 'violating' states or trigger military intervention. This was the case during both the first and second Iraq Wars when suspicions of bioweapons were present. During the ongoing war in Ukraine, Russia, referencing the BWC, has accused the USA of developing bioweapons in Ukraine and has appealed to the UN. Russia also raised the issue at the ninth Review Conference in 2022.[82, 83]

At this last conference, the problem of the lack of monitoring capabilities was raised again, although the USA showed itself somewhat more positive towards changes.[82] A breakthrough on this point would undoubtedly strengthen the BWC. Nevertheless, it must be permissible to express a degree of pessimism here. The forces in the USA that have been particularly negative towards a mandatory control mechanism in the BWC – 'Big Pharma' and the biotechnological industry – have not lost influence in any way; rather, the opposite is true. The protection of industrial secrets is considered more important now than ever, given the significant commercial interests involved. Recently, there have also been several cases of 'hacking' of American pharmaceutical companies, which has not increased the industry's enthusiasm for greater transparency.

A factor that complicates the problem of monitoring considerably is that the text of the BWC is very vague on a crucial point – the distinction between relevant activities for civilian, legitimate purposes and activities aimed at developing bioweapons, or as stated in the text: 'which are not justified for protective or peaceful purposes'. This reflects one of the significant problems in modern research – the dilemma associated with what is called 'dual use' activities. This refers to research that may have a clear, legitimate utility but could also easily be misused for destructive purposes, such as the development of bioweapons. This is very relevant in modern biotechnology, where seemingly 'peaceful' research can easily camouflage 'malicious' activities. The USA's main argument against strengthening the monitoring capabilities in the BWC in 2001 – that reliable monitoring is practically impossible – was, therefore, not entirely unfounded.

Some states' reluctance to increased control from the BWC may also be due to aspects of their alleged bio-defence programmes that they do not want transparency regarding. This could be dictated by security concerns, but illegal bioweapons development cannot be ruled out either. All experience has shown that disinformation and lies always accompany the bioweapons debate.

The BWC was the first international treaty to prohibit an entire category of weapons of mass destruction. It was not until 1993 that a corresponding agreement to prohibit chemical weapons – the Chemical Weapons Convention (CWC) – was reached.[81] In contrast to the BWC, the CWC has built-in clear provisions for mandatory monitoring and control, and it has solid resources to follow this up. The CWC has a permanent staff of 500, while the BWC's permanent staff has been three people (four as of 2022). Unlike the CWC, the BWC also does not have any permanent scientific advisory group. The CWC's annual budget is over $80 million, while the BWC operates on one of $1.8 million, about the size of the budget for a single McDonald's restaurant.[83] Member states are expected to pay a modest annual fee, but some have not done so for several years.[84]

The striking difference in commitments to combating biological versus chemical weapons may perhaps stem from an underestimation of the current threat posed by bioweapons. Other factors, as discussed above, may also play a role. Regardless, the lack of prioritisation of the BWC's work in the coming years could prove to be a fatal miscalculation with very serious consequences.[81, 84]

In addition to the BWC, there are also several other international agreements among groups of states aimed at preventing biological material, equipment, and information from being used for the production of bioweapons. However, the effectiveness of these measures is likely limited.[85]

Chapter 10

A Tool for Terror and Crime – Bioterrorism and 'Ordinary' Crime

The use of biological weapons in warfare has roots that go back to antiquity. However, such weapons can also be deployed in other contexts – during acts of terrorism and in 'regular' criminal acts. The distinction between these latter two categories can sometimes be somewhat fluid. All forms of terrorism are naturally criminal. Both bioterrorism and biocrime are, in contrast to the use of biological weapons in war, modern phenomena that emerged after the bacteriological revolution.

Bioterrorism

The term terrorism is not easy to define. A number of different definitions have been proposed, which emphasize various aspects of terrorism.[21, 86] In this context, a not overly strict definition that covers the many forms of terrorism we have seen in recent decades is appropriate.

In the following, terrorism refers to a group or an individual threatening or using bioweapons to kill or harm civilians, animals, or plants based on political, religious, or ideological motives. Hence, the motives for terrorist acts can vary. The very term terrorism is derived from *terror* – the Latin term for fear. Lenin once stated, 'The point of terrorism is to terrorise.' Terrorist acts undoubtedly have an ability to create fear within a population, but other motives are often more important – political, ideological, religious, or simply revenge. In some cases, the primary intention of bioterrorists is to kill as many people as possible rather than to achieve political goals. In many cases, the motives are mixed.[86]

Bioterrorism has almost always been directed against people, yet there are isolated cases where bioweapons have been aimed at animals. It is also possible that bioterror could be directed against plants, for instance, grain production. Several state bioweapon programmes have, as we have already seen, developed weapons with this in mind.

Most terrorist acts have been committed with conventional weapons. In bioterrorism, bioweapons based on microbes or toxins are used. These can be disseminated in the same ways as biological weapons in warfare. To date, relatively simple methods, such as dispersal via food or water supplies, have been most common in bioterrorism. Aerosol dissemination, which is the most effective method of using bioweapons, has so far been used little by terrorists. The reason is that it is technically challenging to produce effective aerosols of microbes and toxins.

So far, bacteria have primarily been used in terrorist attacks. Terrorists have acquired bacteria either by purchasing them from commercial suppliers, of which there are many, or via theft from microbiological laboratories, committed by individuals with access to those facilities. The toxin ricin is popular among terrorists. This can be produced relatively easily from castor beans in a garage or kitchen.

It is, however, a disturbing fact that more and more individuals have acquired knowledge in microbiology and biotechnology that can be exploited by terrorist groups with relatively modest resources to produce microbes and toxins as weapons. Such expertise can today also be used to manipulate microbes to enhance their pathogenicity. Previously, only state actors had the necessary expertise and resources, as seen in the Soviet Union's bioweapons programme. This development raises concerns that biological terrorism will become an increasing problem in the years to come, involving increasingly dangerous microbes and more sophisticated dissemination methods.

The following examples of bioterrorism illustrate the variation in motives and the chilling possibilities that exist when terrorists use biological rather than conventional weapons.

R.I.S.E. – An Early Attempt at Eco-terrorism

In January 1972, the police in Chicago arrested two teenagers believed to belong to a terrorist organisation called R.I.S.E. This organisation allegedly planned mass murder by contaminating Chicago's water supply with the typhoid bacterium *Salmonella typhi*.[87]

The two arrested, 19-year-old Allen Charles Schwandner and 18-year-old Stephen J. Pera, had become friends in 1971 as students at a college, where they made a half-hearted and short-lived academic effort. They believed they had more important things to do.

Schwandner had a somewhat troubled background with adjustment difficulties and a stay in a psychiatric hospital. Pera was highly interested in microbiology but was almost self-taught in the field. Several who knew him considered him to be very intelligent.

The two founded R.I.S.E. in November 1971. It is not clear what the acronym stands for, but there is some evidence suggesting that R stands for Reconstruction, S for Society, and E for Extermination. The group's ideology was formulated by Schwandner, who wrote a six-page 'manifesto'. This document has not been preserved, but we know some of its content from later trials. According to this manifesto, humanity was destroying nature and itself. To avert a catastrophe, it was necessary to exterminate the entire human race except for a small chosen group that would live in harmony with nature. From this group with the right insight, the Earth would be repopulated. Schwandner wanted the first chosen group to consist of about 16 people of both genders.

One can therefore say that the core of the group's ideology was extreme ecological attitudes, and that R.I.S.E. represented eco-terrorism.

Schwandner and Pera agreed that the necessary extermination of humanity could best be accomplished through global epidemics that could eliminate the entire human race with the exception of R.I.S.E. members, who would need to be vaccinated against the relevant microbes in advance. Schwandner's initial plan was to disseminate the microbes to be used via aerosol spraying over major cities. It was hoped that air currents would carry them around the world. A variety of different microbes were to be utilised. In this way, the many different epidemics that would arise would confuse any investigation of the outbreaks until it was too late.

In his apartment, Schwandner set up a special 'war room'. Here, there were maps of locations where the attacks would be carried out and others showing the likely air currents that would carry the microbes around the world.

Pera was the only one in the group with any understanding of the natural sciences and microbiology. He now sought several ways to increase his knowledge so he could cultivate the microbes to be used in the grand project. Previously, he had partially attended some biology courses and had gained access to a microbiological laboratory at a Chicago hospital where he worked without pay. He now took advantage of this contact by studying microbiological literature and gaining practical experience. Furthermore, he set about acquiring suitable bacteria for the large project. He contacted the microbiological department of a large hospital and requested samples of several bacteria. He actually obtained both *Salmonella typhi* – the typhoid bacterium – and *Neisseria meningitidis* – a bacterium that causes meningitis. By unknown means, Pera also managed to obtain the diphtheria bacterium (*Corynebacterium diphtheriae*), dysentery bacteria (*Shigella sonnei*), and the bacterium that produces botulinum toxin (*Clostridium botulinum*). It is uncertain whether he also managed to acquire *Bacillus anthracis* – the anthrax bacterium. There is suspicion that he was also interested in cholera and plague bacteria, but he likely did not obtain those.

Pera secretly began to cultivate bacteria in the laboratory where he had acquired access. He received practical advice from laboratory staff, who were unaware of his intentions.

However, starting in January 1972, the organisation R.I.S.E. began to fall apart. Several newly recruited members were horrified by the extensive plans for bioterrorism and contacted both local police and the FBI. At the same time, Pera's dubious work in cultivating bacteria was uncovered. The bacterial cultures were confiscated, and Pera was denied access to the hospital laboratory where he had gained entry.

But Schwandner and Pera did not lose heart. Pera obtained new samples of meningitis bacteria and possibly *Clostridium botulinum*, which he began to cultivate in the laboratory at the college he had attended previously. The bioterror plans were now scaled back somewhat from the original goal of striking the entire world with biological weapons. New members were informed that the attacks would target five states in the Midwest. This would occur partly through the dissemination of aerosols in buildings and supermarkets, and partly through the contamination of the water supply. Schwandner and Pera began mapping the water supply in Chicago.

Investigations into R.I.S.E. progressed quickly. On 17 January 1972, the two founders were arrested and subsequently brought to trial for attempted bioterrorism. The prosecution was clearly somewhat unsure of the seriousness of the two teenagers' conspiracy, but chose to respond vigorously. The defence, on the other hand, attempted to downplay the situation, arguing that the plans were far too amateurish to have serious consequences. It remains unclear which bacteria Schwandner and Pera actually had access to and what quantities they had cultivated. One source claimed they had 30–40 kilograms of *Salmonella typhi*, but this is doubtful.

Schwandner and Pera were released on bail in March 1972. They fled to Jamaica, where they hijacked a sightseeing plane and flew to Cuba. They likely expected to be received as revolutionary heroes, but were actually not well received there. Schwandner died in 1974, possibly in prison. Pera was allowed to return to Chicago in early 1975 and reported to the police. After he confessed, he was sentenced to five years in suspended prison.

R.I.S.E. is interesting as an early example of eco-terrorism – a phenomenon that is gaining increasing significance today. Several ideologists within the eco-movement argue for the legitimacy of attacks on buildings and facilities deemed legitimate targets. Although humans have not been regarded as primary targets thus far, some advocates for these ideas have stated that lives might be lost during such actions.

Teenager Pera's impressive ability to acquire dangerous microbes and gain access to microbiological laboratories also illustrates a problem with bioterrorism that is highly relevant today.

'The New Human' and Bioterror – The Rajneeshi Sect

People in the West have often found themselves fascinated by what is perceived as Eastern wisdom, especially when conveyed by charismatic gurus. This also applies to the Rajneesh movement that emerged in India in the 1970s.[52,88] The founder was an Indian philosophy professor, Bhagwan Shree Rajneesh, regarded by colleagues as very intelligent with extensive knowledge in both religious studies and philosophy. Rajneesh eventually rejected all established religions and developed his own religiously inspired ideology. This was based on elements sourced from many religions, particularly Hinduism and Buddhism, as well as Western philosophy and psychology. Conventional morality was rejected as a sign of unfreedom. Central to his teaching was the veneration of sexual activity as a means of spiritual liberation and development, as well as so-called dynamic meditation that included screaming, jumping, singing, and dancing.

Rajneesh proclaimed that through his teachings he would create 'the new human', based on the most valuable aspects of both Western and Eastern traditions. He was extremely charismatic and quickly attracted many disciples, initially from India.[89] He founded a monastery-like centre – an ashram – for the movement in Pune, India. Over time, large numbers of members from the USA and Europe flocked there. Many of the Western members were both very wealthy and well educated. Sect members were called sannyasin, wore reddish clothing, and took Indian names. The sect had significant income, among other things, from many efficiently run business enterprises, commercially driven treatment groups, and sales of the guru's books. There were also donations from wealthy members.

The guru gradually came to be referred to as Bhagwan, which means 'the Blessed One', implying divinity. Although the Hindi term 'sannyasin' is associated with asceticism, this was not part of the Rajneeshi sect's teachings. Bhagwan himself had an unusual penchant for Rolls-Royce limousines, of which he eventually acquired 93. He was almost worshipped as a deity by the sannyasin. When he spoke to them, he sat on a throne-like chair.

From India to Oregon

Bhagwan and his disciples eventually found themselves in increasing conflict with local authorities, partly due to disputes with religious leaders. In 1981, he therefore chose to emigrate to the USA and establish his ashram there. The sect purchased a large property, Big Muddy Ranch in Wasco County, Oregon, where the nearest city was The Dalles, with a population of 10,000.

On their newly acquired property, the sannyasin built a small town named Rajneeshpuram.[89] It eventually gained official status with an armed police force of 60 men, a hospital, a pharmacy, and a hotel. Bhagwan also had his own armed guards, while the police in Rajneeshpuram had access to official police databases. The town had its own airstrip with several planes and helicopters.

There was no doubt that guru Bhagwan was the undisputed leader of Rajneeshpuram, which eventually had at least 4,000 inhabitants. However, he did not take part in the daily management but withdrew into a self-imposed silence. It was his secretary, Ma Anand Sheela, who practically administered the sect and represented it to the outside world with Bhagwan's blessing and regular communication with him. She was assisted by a few loyal women. Sheela's decisions were final, and she tolerated no dissent from the other sannyasin.

The sect had grand expansion plans and soon came into conflict with both the local population and Oregon's strict land development laws. The sect had a team of skilled lawyers who filed a stream of complaints and accusations against both individuals and local authorities. In this conflict, the sect members took a very aggressive stance, making veiled threats in the media, articulated by Sheela among others.

When the tactic of complaints and threats did not succeed, Sheela and her allies decided that they had to take steps to gain political control over Wasco County – if necessary using illegal means. In November 1984, there were elections for the members of the political leadership in Wasco County, with a total of 15,000 eligible voters. Sheela now launched her own candidates.

Oregon has very liberal election laws that the sect attempted to exploit. They brought in a total of 3,000 homeless individuals and street people to Rajneeshpuram from New York and other cities to increase the sect's influence in the election. However, the authorities managed to block this manoeuvre.

Biological weapons as a political tool

In consultation with one of her loyal allies, nurse Ma Anand Puja, Sheela then chose to take more drastic action. They decided to incapacitate the voters in The Dalles by infecting them. There is reason to assume that Sheela was now desperate because she felt her position in the sect was threatened by certain other members who were also in direct contact with the guru. This especially concerned a very wealthy group with a background in Hollywood. Sheela therefore had to achieve results to satisfy Bhagwan.

Puja, Sheela's accomplice, held a central position in the sect's hospital where there was a laboratory she led. She was reportedly very interested in pathogenic microbes and toxins, and it later emerged she had ordered several different bacteria to the laboratory from official suppliers, even though these were not necessary for the hospital laboratory's routines. This was possible because she represented a hospital.

Puja was not well liked among her colleagues at the hospital or among several other sannyasin. Many found her creepy, and she got the nickname 'Sister Mengele', referring to Josef Mengele who conducted horrific experiments on prisoners in Auschwitz. She was also very interested in the HIV virus and AIDS and reportedly had plans to cultivate the virus. This exceeded her expertise, but there were rumours that she had infected certain individuals in Rajneeshpuram with HIV by injecting contaminated blood.

As a weapon in this conflict, Puja chose, among several options, the bacterium *Salmonella typhimurium* and cultivated larger amounts of it in a secret laboratory on the outskirts of Rajneeshpuram. In the middle of summer 1984, preliminary tests were conducted with bacteria-laden material smeared on door handles and urinals in The Dalles, but the results were disappointing.

In September, the main attack was launched. Ten to twelve chosen sect members, 'armed' with test tubes containing Salmonella bacteria, dripped the contents into salads and salad dressings at ten restaurants in The Dalles in multiple rounds. Puja herself participated, masked.

Infection with *Salmonella typhimurium* causes acute intestinal infection with diarrhoea, nausea, and vomiting. The infection can be severe, sometimes fatal – partly due to significant fluid loss, and partly because the bacteria can spread from the intestines and attack other organs, such as the heart. Young children and elderly people with other illnesses are particularly vulnerable to serious disease.

From mid-September, the first infections began to appear following the attack. The number of cases quickly rose to 751. Forty-five of the patients were so severely affected they had to be hospitalised in The Dalles, which had only 105 beds. No one died, and a contributing factor to this was likely that the strain of bacteria Puja had chosen was unusually easy to treat with antibiotics.

A Failed Investigation

Salmonella typhimurium is a common cause of food-borne infections. After the bacterium was detected in the outbreak in The Dalles, efforts focused on determining the source of the epidemic. Both local health personnel and specialists from the CDC (Centers for Disease Control) in Atlanta began extensive detective work. It turned out that most of those who became ill had eaten at certain restaurants in The Dalles, and the 'disease detectives' naturally assumed that the infection originated from food at these restaurants. However, the striking thing was that they could not identify a single source of contamination, which is usually the case in such outbreaks. Ultimately, they concluded that the epidemic was due to poor personal hygiene – specifically inadequate handwashing after using the restroom by staff at the affected restaurants.

However, many in the local population were convinced that the red-clad sannyasin in Rajneeshpuram were behind the outbreak. This included two public employees who had toured the sect's headquarters prior to the epidemic and became very ill after being served a glass of water there. But the suspicions of the local population were entirely dismissed by CDC experts as expressions of provincial prejudice and xenophobia.

The issue remained until a bomb detonated in September 1985. After years of silence, Bhagwan held a press conference. A few days earlier, Sheela and her closest allies had fled to West Germany. The guru now condemned Sheela in strong terms, accusing her and what he called her 'fascist gang' of a range of criminal acts, including attempts to murder sect members and public officials, embezzling the sect's funds, and attempts to contaminate the water supply in The Dalles. He invited an investigation into all of this.

The investigation shed light on the bioweapons attack that Sheela and her allies had initiated. In Rajneeshpuram, documents were found that showed Puja, in addition to *Salmonella typhimurium* used in the attack, had obtained several other more dangerous bacteria, including *Salmonella typhi*, which causes typhoid fever. These bacteria were not found, but a sample of *Salmonella typhimurium* was discovered, which turned out to be of exactly the same type as the bacteria from the major outbreak in The Dalles.

Several prominent members of the sect who had been involved in the bioterror plans chose to testify. They could share that Sheela had led them to understand that Bhagwan himself also knew of and accepted the plans – but this was based on a somewhat unclear audio recording. It turned out that advanced wiretapping equipment had been placed in a large number of locations in Rajneeshpuram, both with the guru and in several private homes and communal facilities.

The investigation further revealed that there had also been attempts to infect the water supply system in The Dalles. Puja had also considered placing dead beavers in the water, as virtually all beavers in the area were infected with the parasite *Giardia lamblia*, which can cause epidemics of intestinal infection.

Finally, Sheela and her allies had a list of various individuals they wished to harm or kill, including a state prosecutor. They went so far as to purchase firearms for this purpose, but an assassination attempt failed.

The results of the investigation led to Sheela and Puja being extradited from West Germany and brought to trial. They confessed to most of the charges and received 20-year prison sentences each. However, they served less than four years due to skilled lawyers. Sheela ended up in Switzerland, where she later ran a home for the elderly and demented.

It is part of the story that the guru himself fled with one of his planes and several of his associates a month after his sensational press conference. He was, however, arrested during a layover in North Carolina and charged with violations of immigration laws. At the time of his arrest, he had 21 suitcases that included 35 jewel-encrusted wristwatches and 17 pairs of designer glasses. The disciples had also managed to bring the guru's throne onto the plane.

Bhagwan received ten years of probation and a $400,000 fine. He returned to India, where he restored his ashram in Pune, still with a large number of sannyasin. He later took the Buddhist name Osho. He died in 1990.

It remains somewhat unclear what Bhagwan actually knew about the bioterror attack in 1984. Was it prudent to make Sheela the scapegoat for this?

The Rajneeshi movement that Bhagwan started still exists, with 175 centres in over 60 countries. His teachings continue to have a significant appeal to many, and they are studied with interest by scholars of religion.

Learning Points from Oregon

The bioterror attack in Oregon in 1984 has several interesting aspects.[89] Firstly, it illustrates that genuinely motivated terrorists can achieve a lot even with relatively unsophisticated use of bioweapons. A high number of people were infected during the attack. If Puja had chosen to use the typhoid bacterium – *Salmonella typhi* – or any of the other bacteria she had acquired, it would likely have resulted in numerous deaths.

The Oregon attack also reveals a characteristic of using bioweapons that differentiates it from other types of weapons of mass destruction, namely chemical and nuclear weapons. When bioweapons are used, it is often challenging to determine if the ensuing infections are the result of a natural outbreak or instigated by human intervention. For bioterrorists, this is often an advantage. That was how the sannyasin assessed it in Oregon.

Lastly, the Rajneeshi sect's attack is an example that the motives for bioterror can be mixed. In this case, sect members had a clear political goal, namely to hinder voter participation from opponents by inflicting illness upon them. At the same time, the underlying motive was the belief in Bhagwan's proclamation that the sannyasin were representatives of 'the new human', and a conviction that they were superior individuals compared to the provincial, prejudiced local populace.

The CDC experts who conducted the 'detective work' following the Salmonella outbreak wrote an article on it intended for a scientific journal. However, at the request of the FBI, the article was not published. The reason was allegedly that it could potentially inspire copycats – new terror attacks inspired by the events in Oregon.

Interest in Bhagwan and the Rajneeshi movement surged after the streaming service Netflix launched the documentary series *Wild, Wild Country* about the subject in 2018. A weakness of this series is that it scarcely addresses the bioterror aspects. After all, the attack in The Dalles is one of the most extensive such attacks the world has ever seen.

Minnesota Patriots Council – Government Hatred of Disgruntled Men

In 1970, Frank Nelson – a retired US Air Force colonel – founded a group called the Minnesota Patriots Council. This movement, which established many chapters throughout the state of Minnesota, believed that the US government was corrupt and that its laws and tax policies were illegitimate. The members claimed that taxation was unjust and aimed at keeping the 'average' citizen in poverty while favouring minorities and socially 'undesirable' individuals. In addition to this political viewpoint, the group held typical right-wing positions on abortion and gun legislation. Arguments were often drawn from both the Bible and the US Constitution.[90]

The movement's strong criticism of Minnesota's tax policy was not entirely unfounded. Throughout the 1980s, farmers and rural populations in Minnesota were severely affected by an agricultural crisis. Thousands of small farmers had to abandon their farms and rightly complained about the government's arrogant and ruthless treatment. This led to conspiracy theories suggesting that the government, in collusion with international forces, aimed to buy up vast areas of land in the state.

Some of those drawn into the Minnesota Patriots Council had valid reasons to be dissatisfied, and there is no doubt that many had personal, familial, and social problems. A small group of around eight militant individuals from the Patriots Council met regularly at the home of one of them, LeRoy Charles Wheeler, in the small town of Alexandria. Several in the group had a military background. The meetings, held on Sundays, were secret. Participants discussed various conspiracy theories and listened to audio recordings of speeches by prominent leaders from the right. The ideological leader or 'brain' of the group was Richard John Oelrich, who had been in a heated conflict with the tax authorities that culminated in their armed representatives and federal police forcing their way into his home. Thus, he had a strong motive for vengeance against the government.

In the spring of 1991, Oelrich and another group member, Dennis Brett Henderson, read an advertisement for a reasonably priced 'assassination kit' named Silent Tool of Justice in a right-wing publication. This could be ordered from a well-known far-right extremist and was based on the toxin ricin. A delivery included both castor beans and an instruction manual explaining how to extract ricin from such beans at home in the kitchen or garage. The advertisement emphasized as a recommendation that ricin was used by both the CIA and the KGB. At the end of the advertisement, it stated that the supplier took no responsibility for misuse of the equipment.

After some discussion, Oelrich and Henderson decided to order the equipment. Henderson took on the task of producing ricin according to the recipe provided. The result was a whitish powder that he put into a can that had contained baby food previously. This was placed in another box on a shelf in a shed at Wheeler's home.

In the following months, the small group had a series of discussions about their newly acquired ricin and how they might use it. Possible targets included government employees, particularly officials from the tax agency, and police personnel in the area. The group was fully aware that ricin could be deadly. They discussed several methods of use. Henderson believed that ricin would penetrate the victim's skin if mixed with certain other chemicals and planned to smear the mixture on doorknobs or in the victim's shoes. He also wanted to inject ricin powder into a victim's car so that it would be inhaled. This would likely have been effective, as inhalation of ricin leads to a quick death. However, the effect of this toxin on skin is rather dubious, as the substance likely does not penetrate intact skin. But if the victim's hands carried ricin from their skin to food, lethal poisoning could nevertheless result.

One of the reasons the group chose ricin as a weapon was that they believed it would be impossible to detect the toxin in the victims' bodies. This is not entirely true, as special methods can indeed demonstrate that a death resulted from ricin poisoning.

Despite the group's imaginative plans for ricin attacks, this was never realised. As early as 1991, one of the group members contacted the police and informed them that some of his friends had plans for violent attacks on police personnel. However, the police felt that the situation seemed rather loose and did not take action. In 1992, the spouse of one of the conspirators, Doug Baker, approached the police. She was in the process of leaving her husband, whom she claimed had threatened to kill her. She told them that her husband had weapons, explosives, and a type of poison. The 'poison' was ricin, which Henderson had produced previously and placed in a can at Wheeler's home. This can was now in the hands of the police.

The police sent the suspicious substance to Fort Detrick, where analyses showed that it contained ricin in a quantity sufficient to kill 129 people.

This discovery was enough to charge Baker and Wheeler with possession of a toxin intended for use as a weapon. Wheeler was charged because his fingerprints were found on the can. The charges against the two were based on a relatively new law, the Biological Weapons Anti-Terrorism Act, which had been enacted in 1989. This law was unknown to the conspirators from the Minnesota Patriots Council and had not been used previously.

Neither Baker nor Wheeler admitted that they intended to use ricin as a weapon, but both were sentenced in 1995 to two years and nine months in prison. Baker was released the following year. The other members of the group had gone into hiding when Baker and Wheeler were arrested. However, in 1995, Henderson and Oelrich were also arrested, charged, and sentenced to 48 and 37 months in prison, respectively.

The case against the members of the Minnesota Patriots Council shows how easy it is to acquire a deadly toxin like ricin for potential terrorists. The new law under which the four conspirators were convicted was introduced in an attempt to prevent bioterror. It allows for conviction for mere possession of a potential biological weapon, even if it has not been used. This law has proven useful in preventing bioterror and has since been invoked numerous times. The United Kingdom and Russia have introduced similar laws, which are lacking in most countries, including Norway.

Aum Shinrikyo – The Doomsday Cult with Biological Weapons

During rush hour on the morning of 20 March 1995, Tokyo's subway – the busiest such system in the world – was as usual full of passengers heading to work. Among them were five passengers who were about to carry out their horrifying task in the underground. Their mission was to kill as many train passengers as possible.

The five special passengers boarded different trains. Each of them had an umbrella and a small brown package that they placed under their seat. The brown packages contained plastic bags filled with liquid laced with the deadly nerve gas sarin. They had planned for the cars they were in to be at the central Kasumigaseki station at the same time. Above this station were a number of central government offices and many important public offices, including the police headquarters.

A few minutes before the trains arrived at Kasumigaseki station, each of the five passengers pierced the brown package under their seat with the tip of their umbrella and immediately left the station after arriving. Waiting for them in the street above were getaway vehicles.

From the brown packages they had left behind, the sarin-laden liquid began to evaporate, and soon sarin gas filled the cars. Within seconds, passengers nearby began to exhibit symptoms – increasing breathing difficulties, visual problems, nausea, vomiting, and involuntary urination and defecation. Many collapsed with convulsions and paralysis, and some died quickly. Panic erupted as passengers staggered toward the exits at each station. Ambulance teams quickly arrived with stretchers and began carrying out passengers who had collapsed. The most dramatic events occurred at Kasumigaseki station after the five trains arrived. In total, fifteen subway stations were affected.

Thirteen passengers from the subway died from the sarin attack, at least 5,000 were injured, and some suffered permanent damage. Some were left in a coma for extended periods. In one woman, the lenses of her eyes fused with the corneas, leading to the removal of both eyes.

This dramatic attack in Tokyo's underground was carried out by Aum Shinrikyo – a peculiar cult led by a nearly blind guru who called himself Shoko Asahara. It was this attack that brought the cult international attention, but in Japan it had been active for many years.[52, 86, 91]

In hindsight, it became clear that the attack on Tokyo's underground was not the first that the cult had conducted with sarin. In 1993, Aum used sarin against the courthouse in Matsumoto, where a trial was under way against the cult in connection with a property purchase. Since it was believed that the case would not go in Aum's favour, they sought to kill the three judges. They were injured but survived but seven other people were killed, and 144 were injured, some seriously.

A New Religion

The Aum Shinrikyo movement, often translated as 'the Supreme Truth', was started under another name in 1984 by Matsumoto Chizuo, who later took the name Shoko Asahara. He was born with glaucoma, which left him blind in one eye and visually impaired in the other. Initially, he made a living as an acupuncturist and masseur while selling Chinese herbal medicines. He then began to take an interest in mysticism, divination, and the possibility of developing supernatural abilities. He founded a yoga school and began to recruit members to it. Over time, his teachings began to take on a religious character, with elements drawn from Tibetan Buddhism, Hinduism, Christianity, and so-called New Age ideas.

The sect leader now presented himself as both the reincarnation of Christ and a Buddha-like figure who would save the world. He was undoubtedly charismatic and was regarded by followers as almost divine. Several cult members paid significant sums to receive some of the guru's hair, a bit of his bathwater, or his blood to drink during initiation ceremonies, thereby partaking in the guru's spiritual power. In 1989, the movement was officially recognised as a religion in Japan.

Aum Shinrikyo attracted more and more members. Many were well-educated young people, some with backgrounds from prestigious universities, and were seen as an elite movement in Japan. A minority of the members became almost like monks and nuns, severing all ties to family and their former lives. The movement established a long series of centres around Japan, with the headquarters at the foot of Mount Fuji.

Asahara increasingly became preoccupied with the idea that the world's end was imminent. He claimed that during a trip to the Himalayas, he had met learned men who informed him of this. He was also concerned with the concepts of the apocalypse in the Book of Revelation in the New Testament. Asahara believed that followers could be saved during the upcoming Armageddon through faith in him. He preached that this would coincide with a third world war in which the US would attack Japan with nuclear weapons. Over time, he set the date for the apocalypse to 1997. When this occurred, Aum Shinrikyo, as the survivors, would take control of Japan. Asahara therefore appointed a 'shadow government' among his followers with around 20 'ministers' and departments corresponding to Japan's.

The guru was completely autocratic within the sect and made all important decisions. He surrounded himself with a select group of leaders who executed his orders with the help of lower ranks in the movement. Any opposition to Asahara's orders was punished severely, sometimes with death. The sect also displayed an aggressive attitude towards the outside world, particularly towards opponents. They even attempted to murder leaders of competing religious movements.

A prominent feature of Aum Shinrikyo was its focus on weapons of all kinds. This was deemed important in light of the impending Armageddon, but also for the defence of the cult against state and police actions. The sect secured conventional weapons, such as the Russian AK-47 assault rifle, which members planned to mass produce themselves. Furthermore, they sought to acquire nuclear weapons and attempted to purchase them in Russia after the dissolution of the Soviet Union.

The cult members were also markedly interested in more fanciful types of weapons, partly inspired by the science fiction literature that many of them read avidly. Weapons based on laser and microwave technology were considered. A group of Aum members travelled to the Tesla Museum in Belgrade to study what the brilliant inventor Nikola Tesla had described in his writings as a 'seismological' weapon – one that could supposedly trigger earthquakes.

Biological Weapons for Armageddon

Among unconventional weapons, Aum Shinrikyo initially focused on biological and chemical types. Here, we shall especially concentrate on the cult's grand plans for biological weapons. This part of their extensive arsenal has remained overshadowed by their chemical weapons, which they managed to employ to the most damaging effects, such as in the attack in Tokyo's underground.

As mentioned, the imminent Armageddon was central to Asahara's teachings. It appears he believed he could avert this catastrophe if he gained power in Japan. In February 1990, there were to be elections for Japan's parliament. Before the election, Asahara founded a political party with 25 candidates, including himself and most of his closest collaborators. The sect used a significant portion of its resources for an active election campaign.

However, the election ended in a crushing defeat for the cult's party. None of the candidates were elected, and Aum Shinrikyo received extremely negative media coverage. Reports began to emerge regarding the cult's inner life and the brutal treatment of its members. Aum retaliated and murdered a lawyer, his wife, and their 1-year-old son, as the lawyer had filed a lawsuit against them on behalf of all the families who had lost members to the sect.

Asahara now decided to expedite Armageddon by attacking the Japanese society with violent means. He also had paranoid notions that the state would attack and destroy his operations. This spurred the extensive investment in biological and chemical weapons.

Aum Shinrikyo had become very wealthy over time, with a solid financial foundation for its activities. Members made generous donations and, as an officially recognised religion, the cult also enjoyed significant tax benefits. Aum operated extensive businesses, including a chain of IT stores, several restaurants and gyms, its own hospital, and other business ventures. The sect was also involved in the production of illegal drugs, collaborating with the notorious Japanese mafia, known as the Yakuza. Aum had offices in several countries, including the US and Russia. In 1995, the cult's membership was likely around 40,000, with 10,000 in Japan and as many as 30,000 in Russia.

Ironically, it was the cult's 'minister' for the 'Ministry of Health and Welfare' who took responsibility for Aum's biological weapons programme. The first weapon they chose to focus on was botulinum toxin. For unknown reasons – possibly fear of discovery – they opted to produce the toxin-producing bacterium *Clostridium botulinum* from soil where the bacterium is normally found, rather than trying to obtain it from other laboratories or through commercial means. They acquired the bacterium and initiated massive cultivation, resulting in several tons of bacteria-laden liquid. The cult intended to attempt to use his liquid as a weapon through spray systems. No attempt was made to purify the botulinum toxin, which would have provided a far more effective weapon.

A Series of Bioweapon Attacks

Throughout 1990, Aum carried out a series of attacks with its botulinum weapon, which was spread from cars. On one occasion, the primary target was Japan's parliament in Tokyo. Other targets included the port of Yokosuka – the US's primary naval base in the eastern Pacific – and one of Japan's major airports, Narita. In 1993, the cult sprayed botulinum toxin in central Tokyo to create disruption during the Japanese crown prince's wedding.

To the Aum actors' great disappointment, these carefully planned attacks had absolutely no effect and were not even noticed by those around them. Clearly, their botulinum weapon was ineffective and there are many possible reasons for this. There is considerable variation in toxin production among different strains of the botulinum bacterium, and it is possible that Aum did not procure a strain that produced a significant amount. Tests conducted on mice with the toxin were also ineffective. On one occasion, a cult member fell into one of the large cultivation tanks filled with the liquid and nearly drowned – without any signs of botulinum poisoning. It is also possible that their botulinum preparation was not stored correctly, causing the toxin's effect to diminish.

After the disappointment with botulinum toxin, the cult decided in 1992 to focus on *Bacillus anthracis* spores – a well-known weapon that is easier to use. First, the responsible 'minister' tried to steal bacteria from a laboratory. When this failed, he likely contacted a colleague at a laboratory where he had worked previously. This gave him access to a strain of anthrax bacteria. Again, mass production was initiated, resulting in stockpiles of several thousand litres of bacteria-laden liquid. But, similar to the production of botulinum toxin, the cult did not attempt to purify the bacterial material from the growth medium, which was used directly as a weapon through spray systems.

In 1993, Aum conducted several attacks with its newly developed anthrax weapon. From the top of an eight-storey building owned by the sect in Tokyo, they sprayed anthrax into the city. Some nearby birds and pets likely died, but nothing else occurred. There was also no effect from spraying anthrax from a car outside the parliament building and in front of the imperial palace. After the initial failed attacks, they attempted to dry the anthrax liquid into a sort of powder, but again, this proved ineffective. On several occasions, Aum members also accidentally inhaled anthrax material during production without any consequences. What could explain the failure of anthrax as well?[92, 93]

Different strains of anthrax bacteria vary significantly in their pathogenic potential. Subsequent investigations of a sample from Aum's bacteria indicated that it was a low-risk variant primarily used for the production of anthrax vaccine (the so-called Sterne strain). Was the project leader aware of this? It is possible that he thought he could alter the bacterium so that it became pathogenic. This could theoretically occur by combining the genetic traits of the vaccine strain they already had with genes from another vaccine strain, the so-called Pasteur strain. However, whether Aum truly acquired the Pasteur strain is entirely unclear.[92, 93] It is likely that they also lacked the necessary expertise to carry out such genetic modification. In any case, Aum Shinrikyo's anthrax weapon was ineffective. When they became aware

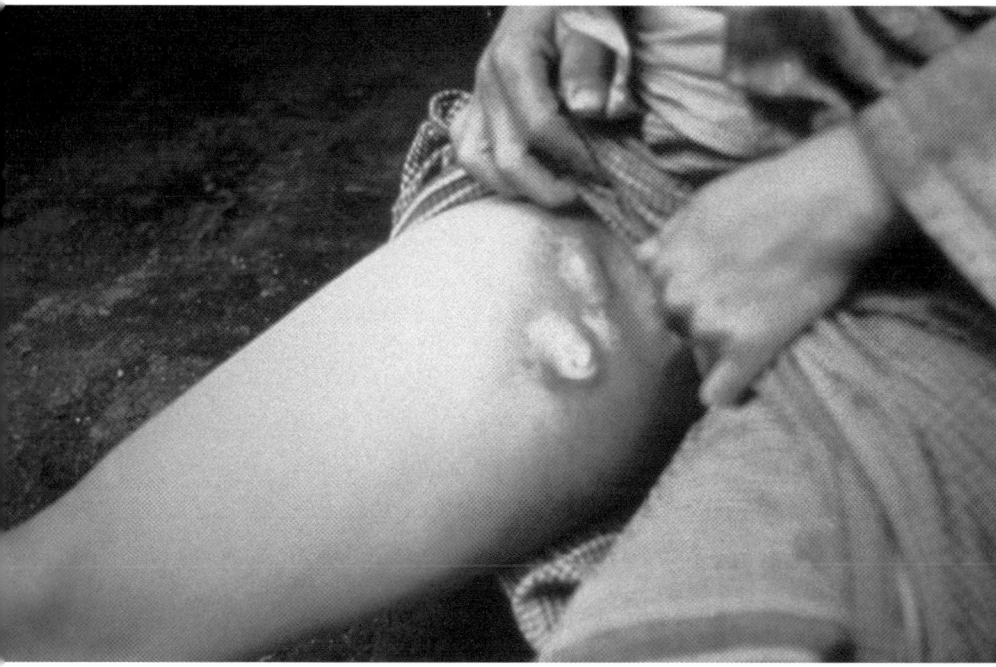

In the classic form of bubonic plague, large, pus-filled lymph nodes (buboes) are seen on the neck, in the armpits, and in the groin, as shown in the picture.

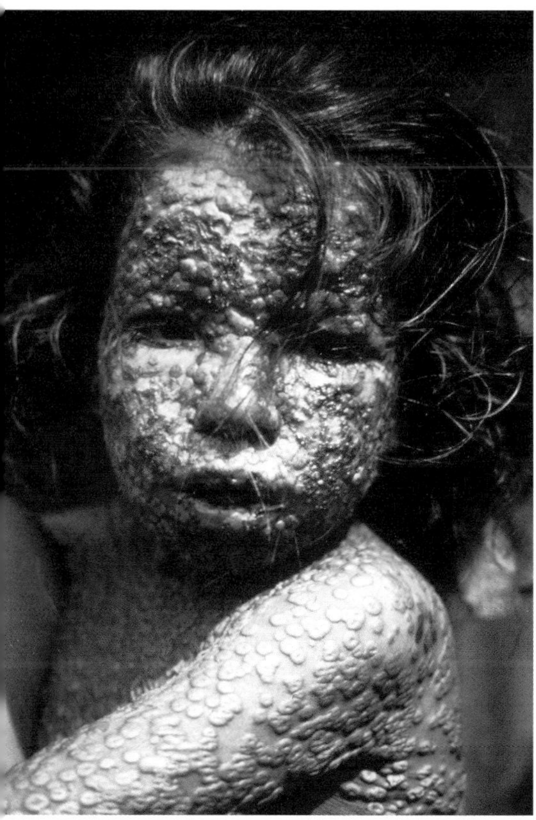

Child with smallpox

The Scythians were outstanding horsemen and archers. They were feared, among other reasons, becaus
they used arrows coated with disease-causing biological material—an early form of biological warfare.

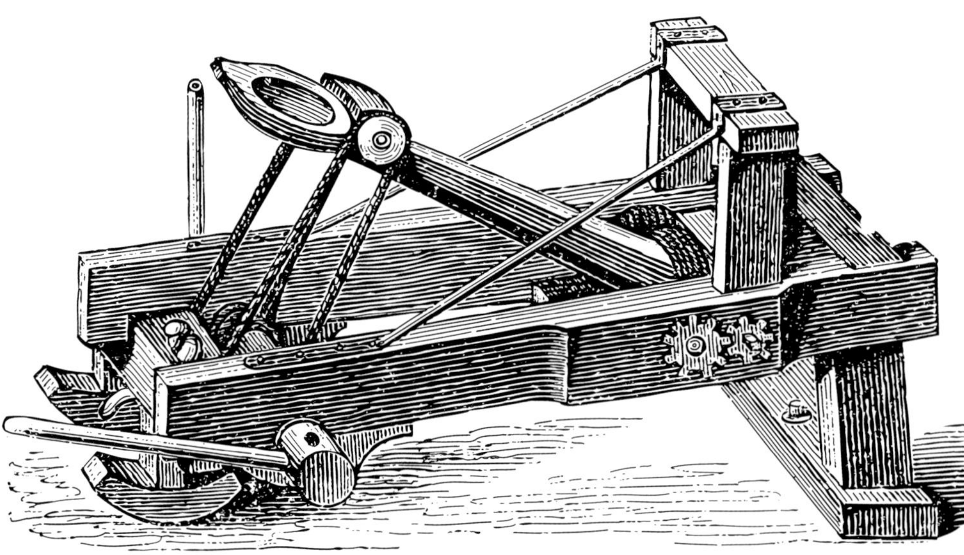

Catapult – a projectile-launching machine used in sieges in antiquity and the. Middle Ages. When th
Mongols besieged the city of Kaffa in Crimea in 1346, they used catapults to hurl plague-infected corpse
into the city.

ıe Seven Years' War from 1756 to 1763 between England and France was also fought in North America. : the end of the war, a number of Native American tribes rebelled against the English in 1763. English ficers then attempted to infect hostile tribes with smallpox by giving them wool blankets and handkerchiefs ed by smallpox patients.

:to von Rosen was a Swedish baron who 'ved in the German army during World War nd conducted sabotage activities in the Arctic ırts of Scandinavia. In 1917, he was arrested in ıutokeino. In his luggage, anthrax and typhus cteria were discovered, which were intended to used as biological weapons against horses and ssibly also reindeer.

General Shiro Ishii led Japan's extensive biological warfare program before and during World War II. He was a physician with training in microbiology and immunology.

Biological weapons experiments on children at Unit 731 in Manchuria.

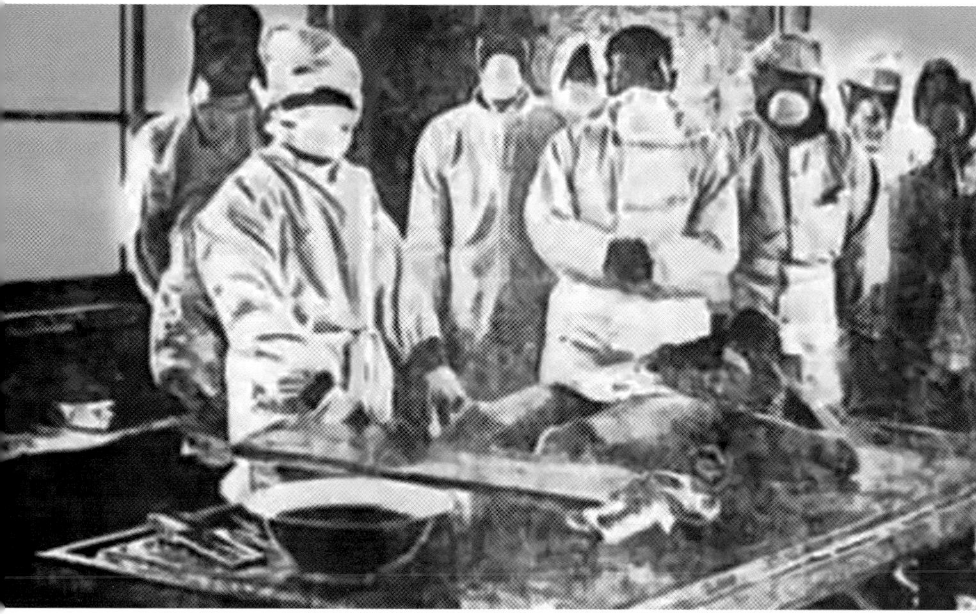

:any of the victims of experiments at Unit 731 were subjected to vivisection—live dissection without esthesia.

ıit 731 in Manchuria was a major center for the development of Japan's biological weapons. The :ility, led by Shiro Ishii, employed 3,000 people and consisted of 150 buildings, including laboratories, :matoriums, and its own airstrip.

The small uninhabited island of Gruinard Island, off the northwest coast of Scotland, was used from 19 to 1943 for open-air testing of anthrax bombs on sheep. After these tests, the soil on the island becar so contaminated with anthrax spores that it was declared a prohibited area for both animals and huma

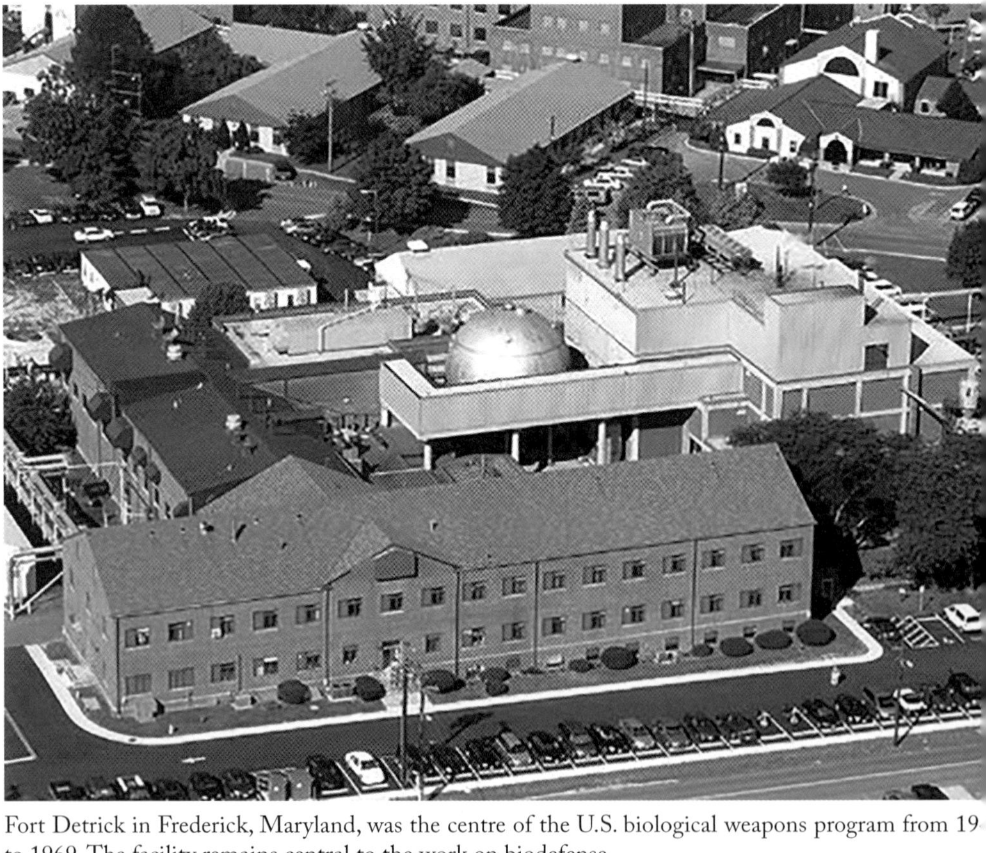

Fort Detrick in Frederick, Maryland, was the centre of the U.S. biological weapons program from 19 to 1969. The facility remains central to the work on biodefense.

Aerial photograph of Vozrozhdeniya Island, located in the Aral Sea on the border between Kazakhstan and Uzbekistan. The island, also known as 'Hell's Island,' was used during the Soviet era for extensive animal testing with various biological weapons.

The remains of the Soviet biological weapons facility on Vozrozhdeniya Island.

In attempts to combat the rebellion against the white regime in Rhodesia during the 1970s, the authorities used specially trained security forces, including their own units of Black, native soldiers – the so-called Selous Scouts.

In a speech to the UN Security Council on February 5, 2003, U.S. Secretary of State Colin Powell claimed that Iraq had biological weapons. As evidence, he presented a test tube that allegedly contained anthrax bacteria.

The so-called mobile biological weapons laboratories that the Americans claimed Iraq had turned out to be ordinary trucks.

David Kelly was one of the United Kingdom's leading biological weapons experts. In 2003, he anonymously claimed in an interview with a BBC journalist that Tony Blair's government had fabricated unreliable intelligence reports to justify the United Kingdom's involvement in the Third Gulf War.

In 1969, negotiations began in Geneva that led to the Biological Weapons Convention (BWC).

The charismatic Indian philosophy professor Bhagwan Shree Rajneesh founded and led the Rajneesh sect, which gained international prominence.
He rejected all established religions and developed his own religiously inspired ideology aimed at creating 'the new human being.'

Shoko Asahara was a nearly blind guru who founded and led the Japanese sect Aum Shinrikyo, which used both biological and chemical weapons to influence society while awaiting Armageddon.

The anthrax letters sent in 2001 to two Democratic senators contained a sheet on which it was written in block letters: 'Death to America. Death to Israel. Allah is Great.' This initially led to suspicion that Al-Qaeda was behind the attack.

Reinhard Heydrich was one of the most prominent leaders in the Third Reich. In 1941, he was appointed Reich Protector (Governor) of Bohemia and Moravia. There, he ruthlessly suppressed all opposition through arrests and executions.

In the 1978 assassination of defected Georgi Markov in London, the Bulgarian secret police used a specially made umbrella that could shoot out a small metal capsule containing the biotoxin ricin. The umbrella was produced by the KGB.

In Mary Shelley's book *Frankenstein*, or the *Modern Prometheus* from 1818, the scientist Frankenstein creates an intelligent monster. The image shows Boris Karloff as the monster in a film adaptation of the book.

A BSL-4 laboratory has the highest level of safety for work with dangerous microbes. In such laboratories, staff must, among other things, work in 'space suits' with their own oxygen supply.

A biohacking laboratory is a simple, unprofessional lab where amateurs can conduct their own research—so-called 'do it yourself' research or biohacking.

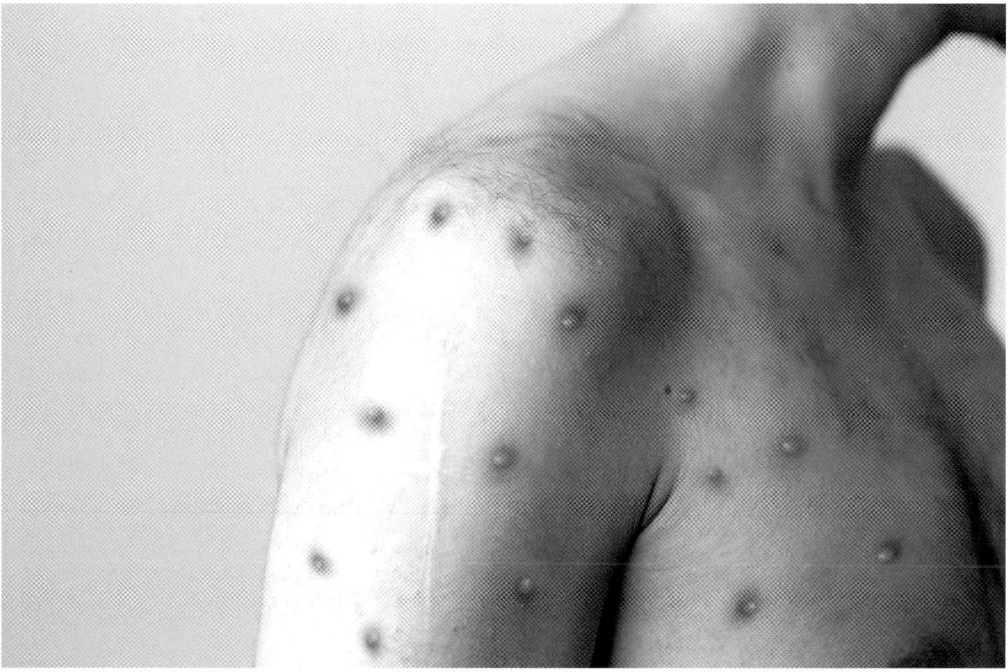

Patient with an outbreak of monkeypox.

Drones will be highly suited as carriers of biological weapons, both in military conflicts and for terrorists.

Prometheus was the rebel who stole the sacred fire from the sky god Zeus and used it to give the spark of life to the humans he had created from clay.

of this, they primarily focused on chemical weapons, particularly sarin, but also several other chemical agents.

We now know that Aum also intended to pursue other biological weapons besides botulinum toxin and anthrax. They worked with *Coxiella burnetii* – the bacterium that causes Q fever. More horrific microbes were also of interest to Asahara and his bioweapons practitioners. In 1992, the guru, as part of an Aum expedition of 40 participants, travelled to Zaire, allegedly to provide aid to victims of the Ebola virus. In reality, the goal was to acquire the virus for the development of biological weapons but it is not known if they succeeded. In any case, no weapon based on the Ebola virus was produced.

Aum Shinrikyo's final attempt at a bioweapon attack occurred just five days before the major sarin operation. Three document folders containing botulinum toxin were to be placed in Tokyo's underground so that the toxin would spread. However, the Aum member responsible experienced pangs of conscience and replaced the toxin with harmless material.

A significant reason for Aum's lack of success with both botulinum toxin and anthrax was that they had chosen the 'wrong' microbes as the basis for their weapons. An important additional factor was undeniably that they lacked the necessary expertise to produce the weapon material suitable for aerosol dissemination. Aum lacked insight into aerosol technology, which would have diminished the effectiveness of their bioweapons even if the microbial material had been the 'right' one. Their home-made spray systems also proved to be ineffective and often failed.[92]

One reason why Asahara resorted to sarin on 20 March was that he had received information that the Japanese police had been planning an operation against the cult for some time. Two days after the attack, Japanese police struck. More than 500 raids were carried out against about 300 Aum properties in Japan. Large quantities of equipment for the production of both biological and chemical weapons were found. Despite this extensive police action, Aum managed to carry out a couple more operations. This time, they used hydrogen cyanide gas – Zyklon B, which was once used in Nazi gas chambers.

More than 300 cult members were arrested following the sarin attack at Tokyo's central station. The legal proceedings took decades. In total, 188 members of Aum were convicted. Six of them received life sentences, while

13 received death sentences, including Asahara. They were all executed in July 2018.[94]

Investigations of the cult were also initiated in other countries, including the US, Germany, Australia, and Russia. In Russia, it was banned, and top leaders were arrested. One of them was the chairman of Russia's National Security Council and a close friend of Boris Yeltsin.

The peculiar and gruesome history of Aum Shinrikyo sparked extensive debate both inside and outside Japan. What led so many young people with higher education from elite institutions in Japan to completely submit to the teachings of guru Asahara and willingly contribute to both individual murders and attempts at mass murder?

There has been discussion as to whether there are specific conditions within post-war Japanese society that led to the alienation of many young people who reacted to the increasing materialism. An intense desire to partake in spiritual values, combined with massive brainwashing from the charismatic cult leader, may have made them willing tools in the cult's assault on Japanese society. The well-known Japanese author Haruki Murakami has argued this in a book based on interviews with cult members.[95]

American psychiatrist Robert Jay Lifton – one of the world's leading experts on the psychological mechanisms of perpetrators and victims in war and political violence – discusses the cult in his book *Destroying the World to Save It: Aum Shinrikyo, Apocalyptic Violence, and the New Global Terrorism.* He writes that in his long career, where he has also interviewed Nazi war criminals, he has never seen such peculiar psychological patterns as those he encountered among the cult members he interviewed. Several of them were doctors from top universities.[94, 96]

A striking aspect of Aum Shinrikyo's history is the Japanese police's remarkable passivity toward the cult over many years. Even when there were strong suspicions – and even clear evidence pointing toward Aum's responsibility for criminal actions – the police failed to intervene. Part of the explanation for this may be a fundamental reluctance to intervene in religious communities. However, there has also been suspicion of corruption within the police as an explanation for this inaction.

Despite Aum Shinrikyo's significant focus on bioweapons, they had no success with their use, as we have seen with the cult's attacks using botulinum toxin and anthrax. Therefore, many have claimed that bioterrorists are unlikely to possess the necessary expertise to carry out large-scale attacks as Aum attempted repeatedly. This is a questionable conclusion.

A significant reason for Aum's lack of success in using bioweapons was that the responsible members – despite high theoretical training in relevant natural sciences – lacked the necessary practical knowledge for the development and use of bioweapons, what is often termed 'tacit knowledge'.[92, 93] They were theorists, not practitioners.

In the nearly thirty years since then, however, there has been a rapid development in biotechnology, with new methods increasingly mastered by more individuals. It is now possible for terrorist groups to recruit individuals with the necessary knowledge to produce several types of bioweapons.

One would think that Aum Shinrikyo would have faded into history after the police actions in 1995. Not only were the leaders arrested, but the cult's properties and financial resources were confiscated, and Aum lost its official status as a religion. However, attempts to ban the entire movement failed for legal reasons. Aum thus continued to exist and rebuild its resources through extensive business activities. They have pursued active and successful recruitment through seminars and propaganda via books and the internet in several languages.

It still appears that the key elements of Aum's 'ideology' live on with a deep reverence for the executed Asahara and belief in an impending Armageddon.[97] From 2000 onward, the movement changed its name to Aleph (the first letter in the Hebrew alphabet). In 2008, the sect split when some members broke away under the leadership of Asahara's successor, Fumihiro Joyu, forming the movement Hikari no Wa – 'Circle of Rainbow Light'. Joyu, who at one time belonged to Asahara's inner circle, received a surprisingly lenient prison sentence.

There is no evidence that sect members have engaged in criminal activity in recent years, but – naturally – they are monitored with great vigilance.

Death is Coming by Mail – the American Anthrax Letters

The USA was still in shock after the attacks on 11 September 2001 on the World Trade Center in New York and the Pentagon when another tragedy struck – also man-made. Letters containing lethal amounts of anthrax spores were sent to individuals in several American cities. Anthrax exposure from these letters led to 22 cases of anthrax infection, and five patients died. The attack prompted a multi-year investigation that is one of the most extensive in American history. This act of terror is often referred to as Amerithrax.[34, 99, 100]

The Outbreak

On the morning of 2 October 2001, a disoriented, clearly ill 63-year-old man named Tom Stevens was admitted to the emergency department at JFK Medical Center in Palm Beach County, Florida. He quickly became comatose. The infectious disease specialist who examined him suspected meningitis, which is inflammation of the protective membranes covering the brain and spinal cord, and performed a spinal tap to examine the cerebrospinal fluid. When he examined the fluid under a microscope, he saw bacteria that he believed could be anthrax. He immediately sent a sample of the cerebrospinal fluid for laboratory testing and received confirmation on 4 October that it was the anthrax bacterium.

The case was made public, and 'disease detectives' from the CDC rushed out to map what was perceived as the natural background for the infection.

There are three forms of natural infection with anthrax: cutaneous infection from direct contact with the bacteria, gastrointestinal infection after ingesting contaminated meat, and pulmonary infection when anthrax spores are inhaled. The latter form is the most dangerous, with a very high lethality rate if antibiotics are not started early. This was the form that Tom Stevens had. He died on 5 October, the day after the diagnosis was made.

This case was very surprising because anthrax had become a very rare infection in the USA. Stevens was only the nineteenth case of pulmonary anthrax since 1900 and the first since 1976. All these cases had occurred in particularly vulnerable occupations but Stevens worked in the editorial offices of a tabloid newspaper in Boca Raton, Florida. The CDC epidemiologists could not find anything in his behaviour and life that could explain the anthrax exposure.

However, it turned out that anthrax spores could be detected both on the keyboard of Stevens' computer and in the mailroom at the editorial office. It became clear that an older man who worked with the mail had been hospitalised on 1 October with a form of pneumonia that was treated with multiple antibiotics. It turned out that he also had an anthrax infection but he survived. Additionally, it could be established that a younger woman who assisted him in the mailroom carried anthrax spores without becoming ill. She reported that she opened a letter that contained a powder of some sort. She got some of it on her dress but discarded the letter, which was never found.

All these findings indicated that this anthrax outbreak was most likely man-made and had no natural background. The FBI launched an investigation, but while there were suspicions of a terrorist attack, officials officially chose a reassuring strategy in communications with the public.

The findings in the editorial offices where Tom Stevens worked raised suspicions that the anthrax exposure had occurred via mail. Confirmation of this came with an interesting discovery in the offices of NBC News in New York. On 12 October, a letter with anthrax spores addressed to the well-known anchor Tom Brokaw was found. The letter had been archived a few days earlier as uninteresting 'hate mail'. The reason this letter was discovered as a carrier of anthrax was that an assistant of Brokaw – a younger woman – had developed a nasty sore on her chest, swollen lymph nodes in her neck, and fever since late September. The doctor she visited believed it could be a spider bite.

However, she did not settle for this explanation. After all the commotion about the anthrax cases in Florida, she had studied pictures on the internet of anthrax infection in the skin and believed that they resembled her own sore. She contacted the FBI, and it ultimately turned out that anthrax bacteria were detected in tissue samples from her wound. A young newly hired woman who helped with mail processing at NBC News in New York had also been on sick leave for a few days due to a sore on her leg, which was also perceived as a possible spider bite and treated with antibiotics. She now remembered that they had received a letter with some powder some time ago. This letter was found in the archive, and it quickly turned out that it contained anthrax spores.

Both the CDC and the FBI now began investigations in the offices of a number of media companies in New York. At the *New York Post*, they found a new anthrax letter. A female employee had recently been treated with antibiotics for a wound perceived as a spider bite. Furthermore, the child of an employee at ABC News had recently been hospitalised with a serious skin infection that also turned out to be anthrax. The assistant to the famous TV anchor Dan Rather at CBS Evening News had a sore on her face, also perceived as a possible spider bite. Anthrax spores were found in the premises.

Now there was no doubt that anthrax spores had been sent by mail to several media companies in various cities. But these were not the only recipients. Two Democratic senators in Washington also received letters.

On the morning of 15 October, a letter with suspicious powder was opened in the office of Democratic Senator Tom Daschle, who was the majority leader in the Senate. The powder was immediately sent for examination, and by the next day it was confirmed: it was anthrax. It turned out that more than 25 employees in the building where Daschle's office was located tested positive for anthrax but showed no signs of illness.

Only on 15 November was an anthrax letter to Democratic Senator Patrick Leahy discovered. The delay was due to the letter being misdirected. Leahy was the chairman of the Senate Judiciary Committee.

Until then, it was not expected that the anthrax letters would pose any significant risk to the postal workers who handled them. This soon turned out to be a misjudgement, and it had tragic consequences.

All the letters were, in fact, mailed from the same place – Trenton in New Jersey – and were processed at the Hamilton post office there. The letters were sent in two batches, on 18 September and 8 October. It was later discovered that one of the postal workers there developed anthrax on the skin, and twelve of the employees tested positive for anthrax without showing any signs of illness.

Now cases of anthrax began to emerge among postal workers. On 19 October, pulmonary anthrax was confirmed in an employee at the large Brentwood post office in Washington, which had 2,000 employees, and in the days following, three more cases of severe lung infection with anthrax were confirmed among his colleagues. They worked at the post office that processed mail for Congress.

A Fatal Misjudgement

On 13 October, the day after anthrax was confirmed in the letter sent to Senator Tom Daschle, several congressional offices were closed for security reasons, and antibiotics against anthrax were offered to prevent infection. It was not until 21 October that this happened at the Brentwood post office after the first serious anthrax cases were detected, and several people had died.

In hindsight, two explanations can be identified for why authorities did not close the affected post offices more quickly, which undoubtedly would have saved lives. First, it was extremely difficult for the management of the US Postal Service to close central post offices unless there were compelling reasons for doing so. But equally important was that health authorities – primarily the CDC – did not understand that there was a significant risk for postal workers. There was a fatal lack of knowledge about critical aspects of anthrax infection and a lack of insight into the conditions at the large post offices where mail was sorted mechanically.[34]

Leading experts claimed that only those who opened a letter containing at least 1–2 grams of anthrax spores were at risk of infection. Based on old data from animal experiments before 1969, they believed that the risk of infection was minimal if fewer than 8,000–10,000 spores were inhaled. Later experiments, however, have shown that inhaling just a few spores can lead to infection in certain individuals. Furthermore, scientists in the Canadian bio-defence programme made very interesting discoveries regarding the risk of infection specifically associated with anthrax letters in 2000. They found, first of all, that anthrax spores in hazardous amounts spread throughout a room after a letter is opened and remained airborne for ten minutes. They also demonstrated that spores can escape from an unopened letter through small openings in the corners of the seal. In short: people can be infected with anthrax from unopened letters.

The Canadian researchers presented their findings at various scientific meetings in the spring of 2001, including in the USA. In March, two of the researchers discussed their views at a meeting at the CDC. It is strange that the CDC did not recognise earlier that postal workers were at risk and act accordingly. The mechanical sorting of letters involves somewhat rough handling. Additionally, these machines are routinely cleaned for dust using compressed air, which creates airflows that increase the spread of anthrax spores. Yet central post offices were closed only several days after the congressional offices, and antibiotics were also offered to postal workers later than to Capitol staff.

In total, the anthrax letters led to 22 cases of infection with symptoms – 11 cases of skin infection and 11 cases of pulmonary anthrax. Five of the latter group died. Several survivors who had pulmonary anthrax faced health problems for years afterward. Apart from two of the cases that resulted in death, all other patients had job tasks that brought them into contact with mail. In reality, thousands were exposed to anthrax spores due to the infected letters, since other letters could 'pick up' spores and carry them forward.

At least 33,000 people believed to have been exposed received preventive treatment with antibiotics, primarily ciprofloxacin – the first-choice treatment for anthrax. This likely saved many lives.

A Faltering Investigation

In the beginning, the authorities tried to downplay the suspicion that it was a terrorist attack but after the discovery of the first anthrax letters, it became impossible to cover up these events. The major question everyone was asking was who was behind it?

Early in the anthrax outbreak, it was confirmed that the bacteria that took Tom Stevens' life in Florida were of a special type – the so-called Ames strain. This strain, which is highly virulent, had originally been found during an anthrax outbreak in cattle in Texas. Since then, it has never been found naturally, but has continually been used by several American laboratories working on vaccines and other projects important for bio-defence. When the Ames strain was identified during this outbreak, it was clear that it was not a naturally occurring infection.[99]

The USA had just suffered the spectacular terrorist attack on 11 September for which Al-Qaeda was responsible. Immediately following 11 September, the American authorities feared that Al-Qaeda would swiftly follow up with a bioterror attack involving anthrax. Osama bin Laden and Islamist terrorists thus came under early scrutiny after the anthrax attack. Even President George W. Bush expressed suspicion regarding this in a speech. One of the terrorists from 11 September had allegedly been treated weeks before the attack for a nasty leg wound described in a way that could have indicated anthrax.

The suspicion that Al-Qaeda was responsible for the anthrax attack was strengthened by the fact that the letter sent to Senator Tom Brokaw contained a sheet of paper on which it was handwritten in block letters: 'Death to America. Death to Israel. Allah is great.' A similar message followed the letter to Senator Tom Daschle.

However, extensive efforts to find anthrax in the remains of the plane that crashed at the Pentagon, including analysis of the terrorists' body parts, yielded negative results. The same was true for investigations of the terrorists' whereabouts in the time leading up to 11 September 11.[100]

The Al-Qaeda leads faded away. This Islamist organisation had indeed expressed interest in bioweapons, including anthrax but many experts doubted that it had the necessary expertise to produce anthrax spores suitable as weapons.

Closer examination of the anthrax spores in the deadly letters provided very interesting results. In particular, the letters to the two senators were found to contain anthrax spores that had clearly been processed in such a way that required considerable expertise. Some experts even argued that the processing was so sophisticated that it could only have been a state actor with the necessary resources behind it. This led to suspicion directed at Iraq, which had demonstrably had an extensive bioweapons programme. However, UNMOVIC, led by Hans Blix, had, as mentioned previously, found no evidence that Iraq still had bioweapons in 2002. The final analyses of the powders in the letters also showed that it was not definitively 'weaponised', meaning that it was not fully completed as a bioweapon.

Over time, suspicions began to shift in a direction that was actually even more shocking than theories of foreign perpetrators behind the attack. Could the perpetrator be an 'insider' in the American bio-defence programme? This included not only the laboratories at Fort Detrick but also several civilian laboratories. The specific Ames strain that caused all the anthrax infections was precisely the anthrax variant being worked on in these laboratories.

At a specific laboratory, highly refined methods had been developed to distinguish between Ames bacteria from different laboratories and hopefully identify which laboratory the specific killer bacterium came from. Investigations showed that the bacterium in question was the so-called RMR-1029 variant of the Ames strain. RMR-1029 existed in just one place – at the bio-defence laboratory at Fort Detrick! The perpetrator or perpetrators must have had ties to this laboratory and had access to the RMR-1029 bacteria.

The first person to come under FBI scrutiny was a doctor with bioweapons expertise named Steven Hatfill. He had indeed stopped working at Fort Detrick more than a year earlier but lived nearby and liked to present himself as a bioweapons expert in the media. The FBI's aggressive investigation of Hatfill ended with him being publicly vilified as the likely perpetrator for an extended period, even by a government attorney. However, the FBI found absolutely no evidence against Hatfill, although this brought immense stress upon him. He subsequently filed a lawsuit against the government, which ended with him receiving a settlement of $5.8 million.

Despite the lack of success in the Hatfill investigation, the FBI remained convinced that the individual responsible for the anthrax letters was an employee at Fort Detrick with access to the RMR-1029 bacteria. Thorough interviews were conducted with a large number of researchers and laboratory technicians but by late 2003 and into 2004, the FBI focused its attention on one of Fort Detrick's leading anthrax experts – Bruce Ivins.[100]

Ivins had worked with anthrax at Fort Detrick since 1980 and had made fundamental new discoveries concerning the bacteria's properties. His findings had proven vital for vaccine development. He was a sociable and well-liked family man but had a history of mental issues that indicated bipolar disorder. The management at Fort Detrick was aware of his mental health problems but believed they did not affect his work.[100]

In the following years, the FBI's investigation continued to focus on Ivins, who increasingly became stressed and depressed. He was prescribed medication for depression while he began drinking. He still worked at Fort Detrick, but from 2007 on, not in the most specialised projects requiring security clearance. He sought legal help under the increasing pressure from the FBI. In July 2008, he committed suicide after realising that charges would be brought against him.[100]

Following Ivins' death, the FBI declared that they considered the case solved, concluding that he was responsible for the anthrax letters in 2001. This conclusion sparked extensive debate, as many strongly doubted Ivins' guilt – including anthrax researchers, politicians, friends, and his closest family. No reasonable motive was ever established.

The prestigious National Research Council was tasked by the FBI in 2009 with reviewing the evidence. Their conclusion was that the scientific findings upon which the FBI built its case were likely consistent with Ivins being guilty, but there was no definitive proof for this, meaning that the question of guilt remained open. In 2009, a court-appointed commission of psychiatrists and psychologists examined the case. Their conclusion strongly supported the FBI's view. The commission believed that Ivins' severe psychiatric problems were of such a nature that they strongly indicated he was the perpetrator.[7]

Many still doubt whether the FBI found the responsible party behind the anthrax attacks in 2001. The theory that Al-Qaeda was behind it still has its supporters.

The Aftermath of Amerithrax

The anthrax letters led to immense fear in the American population. Not only had the USA become the victim of a significant terrorist attack on 11 September by an Islamist organisation; now a new attack had come from a scientist who was supposed to protect the public through his work in bio-defence. For a long time, many Americans actually feared opening letters. Many went out and purchased ciprofloxacin to protect themselves against anthrax.

The FBI's inability to find the perpetrator – which ultimately ended with a conclusion that many perceived as dubious – likely led to diminished trust in the American judicial system.

The anthrax attacks, along with the terrorist actions of 11 September, had far-reaching consequences both within and outside the USA.[34, 100] The practical implementation of decontaminating all the contaminated buildings posed significant problems and took many months. Altogether, this cost about $330 million. Expenses for antibiotics for prevention and treatment are estimated at $135 million.

Far greater were the costs associated with the radical reorganisation of the American bio-defence that occurred as a result of the 2001 terrorist attacks. In late 2003, the Department of Homeland Security was established – a new department that included 22 different government offices and organisations. A main purpose of this new department was to prevent and reduce the consequences of terrorist attacks, but the department also had other duties related to emergencies and catastrophic events.[34]

As we have seen, there was also a significant investment in bio-defence research that received billions of dollars in support annually. Many of these funds were channelled through the National Institute of Allergy and Infectious Diseases (NIAID) to a wide range of civilian research laboratories at American universities and in the private sector. Anthony Fauci – the influential leader of NIAID – who became internationally known during the Covid-19 pandemic, played a central coordinating role in the American bioweapons programme, which also included military laboratories.

The anthrax attacks revealed various shortcomings in the American defence against biological weapons and this led to new laws aimed at combating bioterrorism.[2] In 2001, the so-called Patriot Act came into effect, which limited access to biological materials with potential for weapon development. In 2002, a law was enacted establishing that special control regulations applied to certain microbes and toxins – the so-called select agents.

From 11 September 2001 to the end of 2008, the USA spent $48 billion on its bio-defence. As part of this intensified research, a large number of high-security laboratories for research on particularly dangerous microbes were established. There are four levels of biosafety, and such research should only occur at levels 3 and 4. In 2008, the number of level 4 laboratories, the highest biosafety level, had tripled since before 2001. There were now several hundred level 3 laboratories.

Paradoxically, this increase in high-security laboratories could have actually increased the risk for American citizens. One issue is that any potential turncoats with terrorist intentions in these laboratories may gain access to dangerous microbes, as the FBI believed had occurred in 2001. Even more serious is the risk that such microbes could leak into the environment due to laboratory accidents and potentially trigger epidemics. Such incidents are far more common than many might assume. The fear that this could happen has led, on several occasions, to strong local protests, which have in turn resulted in planned new high-security laboratories not being established.

That a number of civilian research laboratories at universities are now involved in bio-defence research has, according to critics, also led to less transparency regarding ongoing research. This undermines a fundamental ideal within the world of research. As we have seen in other countries' alleged bio-defence programmes, such as the Soviet one, offensive bioweapons research can be conducted in secrecy under the guise of civilian research purposes. This is, of course, easier to carry out when there is not full transparency about the activities. As mentioned previously, the American government, under George W. Bush, succeeded in 2001 in preventing the BWC from being amended to enable more effective monitoring of the bioweapons situation. This attitude persisted in the following years.

Why Haven't There Been More Bioterror Attacks?

Most instances of terrorism since the early 1900s have involved conventional weapons – bombs, dynamite, and firearms. One might then ask: why have we seen only a few instances of bioterror? It is unlikely that moral scruples have deterred terrorist organisations from using biological weapons. Fanatical terrorists often view their fellow citizens as deadly enemies and dehumanise their opponents.

In the few known cases of bioterror, the use of weapons has also been quite primitive, as we saw with the Rajneeshee movement in Oregon where bacteria were spread manually on food items in restaurants. In the other most famous case of bioterror – the Aum Shinrikyo cult – it was also very ineffective, as the members lacked the necessary practical experience with preparing microbes as weapons, and they did not master the technology for effective aerosol dissemination.

There is reason to believe that there is a certain conservatism among many terrorists that leads them to prefer well-tested, conventional weapons like explosives and firearms, and hesitate to utilise bioweapons, which after all require expertise and effort in both preparation and execution of terror attacks.

With this historical background, many have argued that the risk of bioterror is exaggerated and that the fear of such attacks is out of proportion. For example, in the USA, large resources have been allocated to prevent and possibly respond to bioterror attacks. Although the scope of this initiative has been criticised and can certainly be debated, there is no doubt that bioterrorism poses a risk in today's society – including in countries other than the USA.

It is interesting to note that Anders Behring Breivik, during the planning of his terror attack in Norway on 22 July 2011, also considered using anthrax as a weapon.[101] He ultimately chose to focus on firearms and explosives.

The production of highly effective bioweapons capable of killing a very large number of people requires significant financial resources, in addition to expertise, something that only state actors have had thus far. However, this may change in the coming years. More and more people today are receiving education and insight into practical biotechnology. It is also a fact that terrorists, even with relatively modest resources, can produce or otherwise acquire microbes for terrorist purposes – although not initially as sophisticated weapons of mass destruction. However, this too may change in the coming years. If bioterrorists eventually learn to master the technical aspects of high-efficiency aerosol weapons, modern drone technology will make it possible to spread microbes and toxins effectively over larger distances. In that case, a significant number of people could be affected, and biological weapons could also become real weapons of mass destruction in the hands of terrorists.

But terrorists can already cause considerable harm with less-advanced bioweapons that do not satisfy state actors' quality standards regarding durability, aerosolisation, and possibilities for mass production. Bioterrorists can achieve their goals even with moderate death tolls. Just how extensive and paralysing such fear can become was clearly evident in the USA after the anthrax attacks.

There is evidence that there has been interest in bioweapons among terrorist organisations in recent years; this includes Islamist terrorist groups.[102] As mentioned previously, Al-Qaeda came under scrutiny after the anthrax attack in 2001. Representatives of this organisation have repeatedly stated that they wish to invest in bioweapons in their fight against Americans and their allies. Osama bin Laden stated to *Time Magazine* in 1999: 'Acquiring weapons to defend Muslims is a religious duty. If I have obtained these weapons (that is, weapons of mass destruction), I thank Allah for enabling me to do so … It would be a sin for Muslims not to try to acquire such weapons that would prevent the infidels from harming Muslims.'[103] In a so-called fatwa, the Saudi scholar Sheikh Nasir bin Hamid stated in 2003 that Muslims are both religiously and legally entitled to use all available means – including biological weapons – in their fight against the USA.[104]

In 1999, bin Laden allegedly attempted to acquire microbes such as anthrax, Ebola virus, plague, and Salmonella bacteria, as well as botulinum toxin, among others from states in the former Soviet Union. Al-Qaeda is also said to have received assistance from Russian bioweapons experts. Representatives of the terrorist organisation have repeatedly called upon Muslim scientists to contribute to the bioweapons effort and have encouraged their supporters to acquire the necessary education.

There is solid evidence that Al-Qaeda also began work with anthrax as a weapon, possibly also botulinum toxin. After the Americans invaded Afghanistan, they found several laboratories equipped for bioweapon production, and traces of anthrax were also discovered there. Detailed production plans for new bioweapons facilities were in evidence. Al-Qaeda's plans for biological weapons, as is well known, were not realised.[102, 104]

In recent years, Al-Qaeda has been weakened significantly, but still exists and has likely not changed its ideology. ISIL – the Islamic State in Iraq and the Levant – is still active, even though the organisation has faced significant setbacks. ISIL has particularly focused on chemical weapons, however, in 2018, German police in Cologne revealed an Islamist who was planning a terrorist attack using ricin, which he had produced from castor beans according to instructions from a jihadist website. He was in contact with ISIL, which had allegedly proposed this attack.

In the spring of 2023, another Islamist was arrested in Germany, suspected of planning terror attacks using, among other things, ricin that he would produce himself according to a recipe found on the internet.[105] In Norway, a younger man was arrested in 2023 and charged with producing ricin, but the motive for this is currently unknown.

Ricin is very toxic but is not suitable for attacks on large numbers of people. Since it is quite simple to produce following instructions available on the internet, it is probable that we will see more ricin attacks.

Overall, the few bioterror attacks so far should not be taken as a sign that biological weapons are not of interest to terrorists. Many have pointed out that the immense impact the Covid-19 virus had on global society could serve as inspiration for future terrorists. Additionally, there are troubling trends in, among other things, biotechnological research that clearly increases the threat of bioterrorism. We will return to this issue.

Bioweapons and 'Ordinary' Crime

While the use of bioweapons by terrorists is well known, fewer may be aware that such weapons have also been employed extensively in ordinary forms of crime. In these cases, bioweapons have been used intentionally against individuals. The motives have been political or more personal, as seen with other forms of crime.[86]

Political Biocrime

Most known instances of politically motivated bioweapon usage have been carried out by states or possibly by intelligence organisations. The victims have often been politically central figures, but other high-profile individuals have also been targeted.

The Attempted Assassination of Abraham Lincoln

An early case of attempted murder using a bioweapon dates back to the American Civil War from 1861 to 1865. The intended victim was none other than the leader of the Union states, President Abraham Lincoln.[7, 106]

In the summer of 1864, Dr Luke P Blackburn, who was an agent for the Confederacy, planned an assassination attempt against President Lincoln by attempting to infect him with yellow fever.

The viral disease, known as 'Yellow Jack', has significant lethality and wreaked havoc on the eastern coastal cities of the United States during the 1800s. In the most severe cases, the patient often experiences bloody vomiting, known as 'black vomit', due to bleeding in the stomach. At that time, it was believed that transmission occurred through contact with an infected person, especially with 'black vomit'. It was not until the turn of the century that it was shown that yellow fever is caused by a virus transmitted from person to person through the bites of several mosquito species, primarily the *Aedes aegypti*, which is particularly adapted to humans.[13] Blackburn was not aware of this.

Blackburn sought to contaminate clothing with 'black vomit' from yellow fever patients. The clothes were packed in a suitcase that was to be sent to President Lincoln. Blackburn also organised shipments of contaminated clothing to several Northern cities, where they would be sold in the hope of triggering yellow fever epidemics.

Whether Lincoln ever received Blackburn's prepared clothing is unknown. In any case, this alleged bioweapon attack did not lead to yellow fever, since Blackburn lacked insight into the virus's actual mode of transmission. It was a common bullet that ultimately took Lincoln's life at the theatre on 14 April 1865, when the actor John Wilkes Booth shot him in the head from behind while allegedly shouting *'Sic semper tyrannis'* – 'Thus always to tyrants'.

It's worth noting that Blackburn, who sought refuge in Canada, was never convicted for his yellow fever assassination attempts. He later returned to his home state of Kentucky, where he was elected governor in 1879. He vehemently denied claims that he had been involved in yellow fever assassination attempts during the Civil War. Subsequent known political bioweapon assassinations occurred after microbes were discovered as causative agents of infectious diseases. The attacks could therefore be planned in a more rational manner than Blackburn's misguided gamble on yellow fever as a weapon.

During World War II, biological weapons were not used militarily, except for the Japanese army's ruthless warfare in Manchuria and China. However, the warring parties were not strangers to using such weapons for assassinations of prominent individuals among their enemies.

The Assassination Attempt on Hjalmar Schacht

In the early 1940s, the American intelligence organisation OSS (Office of Strategic Services), which was a precursor to the CIA, executed a bioweapon attack against the central German economist and politician Hjalmar Schacht. Schacht had played a significant role in German economic policy for many years, including as head of the Reichsbank from 1923 to 1930 and from 1933 to 1939, as well as Economics Minister from 1934 to 1937. He was regarded as the mastermind behind Nazi Germany's economic policy.

The goal of the attack on Schacht was not to kill him but to prevent him from attending an important economic conference. As a bioweapon, the

OSS used staphylococcal enterotoxin produced by *Staphylococcus aureus.* This toxin was likely added to Schacht's food and caused him nausea, vomiting, diarrhoea, and headaches lasting a couple of days. The assassination was considered successful. This story was first known several decades after the war.[52]

Reinhard Heydrich – The Butcher of Prague

The successful assassination of Reinhard Heydrich, one of the most prominent leaders in the Third Reich, received far greater publicity. He was the head of the Gestapo and Heinrich Himmler's second-in-command in the SS. Heydrich was central to the planning of the genocide of the Jews. In 1941, he was appointed Reich Protector (governor) of Bohemia and Moravia. Here, he cracked down ruthlessly on all opposition with arrests and mass executions. It was not without reason that he was nicknamed 'the Butcher of Prague', 'the Executioner', and 'the Blond Beast'. Hitler referred to him as 'the Man with the Iron Heart' – clearly meant as a compliment.

Heydrich was a prime target for the Allies. Under the code name Operation Anthropoid, the Special Operations Executive (SOE) – a subdivision of the British Secret Service – planned an assassination in cooperation with the Czech resistance movement. This was to be carried out by two Czech resistance fighters trained in England by the SOE and flown into Czechoslovakia.[40, 86]

On the morning of 27 May 1942, Reinhard Heydrich was driven in an open car from his home in one of Prague's suburbs to his office at the Prague Castle. In a turn where the car had to slow down, the two assassins were waiting. As one's machine gun jammed, the other threw an anti-tank grenade at Heydrich's car. Heydrich was not hit directly but was struck in the back by shrapnel. He was quickly transported to a hospital, where he underwent surgery. In the following days, he appeared to recover, but then developed a fever on 2 June and died the next day after falling into a coma.

The German doctors concluded after the autopsy that the cause of death was septicaemia – a systemic bacterial infection.

In the 1970s, a theory emerged suggesting that the true cause of Heydrich's death was botulinum toxin placed in the anti-tank grenade that injured him. This type of grenade had reportedly been modified by the Secret Service so that it allegedly contained test tubes with botulinum toxin secured with tape. A significant reason for the spread of the theory about botulinum toxin was that Paul Fildes – the dynamic head of the British bioweapons programme at Porton Down – once confided to colleagues that he had 'a hand in' Heydrich's death. Fildes had allegedly placed botulinum toxin in the grenade that killed him.

However, the theory that Heydrich's death was due to an attack with botulinum toxin is very doubtful. The cause of death remains unclear. Virtually all documents describing Heydrich's illness period after the assassination have vanished. All the doctors involved are dead, and the autopsy performed was not complete. One of the doctors who led Heydrich's treatment later changed his original assessment and expressed doubt that bacterial sepsis was the cause of death. As far as I know he did not propose any alternative hypothesis.

Regarding Paul Fildes' claim about botulinum toxin, it is worth noting that several colleagues have stated that he privately enjoyed impressing others with boastful stories. At one point, Fildes worked with botulinum toxin at Porton Down but lost interest in this toxin as a weapon. Finally, the modification of the anti-tank grenade used against Heydrich was probably made to lighten it, not to contain botulinum toxin, which would hardly have lasted for several months before use, as was the case here.[40, 86]

Regardless of Heydrich's cause of death, the assassination had devastating consequences for the Czech people. The village of Lidice with 500 inhabitants was obliterated. All the men were shot, while the rest were sent to concentration camps. In the small village of Lezaky with 50 inhabitants, all adults and most children were shot. Additionally, several thousand Czechs were arrested, and over a thousand executed.

Tito – A Thorny Ally

During the Cold War, the Soviet Union built one of the most extensive bioweapon programmes the world has seen. However, there is no solid evidence that the Russians used their bioweapons militarily. Nonetheless, there are examples of bioweapons being attempted for assassinations of individuals.

From 1945 to 1980, Josip Broz Tito was the undisputed leader of the newly established communist republic of Yugoslavia. In the early years, he collaborated with Joseph Stalin and the Soviet Union. Over time, he reacted to the Russians' interference in Yugoslavia, and in 1948, the collaboration was broken. After this, Tito pursued a more independent policy. This was a thorn in Stalin's side, who now attempted to have Tito removed with the help of the intelligence service, still led by Lavrenti Beria. When this failed, Beria decided at the end of 1952 to kill Tito. An experienced agent with the code name Max was assigned this task.[86]

Among several methods, the use of bioweapons was also considered. A plan was formulated where agent Max, during a private meeting with Tito, would trigger a hidden mechanism that released the plague bacterium *Yersinia pestis* in sufficient quantities to kill everyone in the room. Max himself was allegedly to be protected by receiving 'a pre-emptive anti-plague serum'. It is doubtful whether the protection would have been effective. Max was also instructed to write a 'goodbye letter' to his wife in advance. Nonetheless, this plan was never carried out as Stalin died in March 1953.

Solzhenitsyn – An Irritating Author

It was not only political leaders who were subjected to KGB assassination attempts. In the 1960s and '70s, the KGB under its leader Yuri Andropov conducted intensive work to combat regime critics – dissidents – who were deemed a threat to the Soviet Union. One of the most prominent among them was the author Alexander Solzhenitsyn, recipient of the Nobel Prize in 1970. With novels including *One Day in the Life of Ivan Denisovich* and *The Gulag Archipelago*, the world became acutely aware of the dark sides of the Soviet system.

KGB leader Andropov became almost obsessed with Solzhenitsyn. Soviet leader Leonid Brezhnev had decided that the author could not be arrested due to his international fame, therefore the KGB chose to attempt assassination with bioweapons, which could hopefully be carried out more discreetly. The knowledge of this bioweapon assassination comes from both a KGB colonel who participated and from the memoirs of a general in the KGB, Oleg Kalugin.[86]

One day in August 1971, Solzhenitsyn was at a grocery store in the city of Novocherkassk. He recounts that he had been completely healthy that morning but later in the day, he began experiencing severe pain in the skin on the left side of his body. Eventually, large blisters developed over his hips, abdomen, and back, and he felt miserable. It took over three months before he recovered, during which time he could barely get out of bed or write. The doctors could not provide a conclusive diagnosis.

The participating KGB colonel in the three-man team that carried out the assassination recounted that the leader smiled afterwards and said: 'It's over. He's not going to live much longer.' It is uncertain which bioweapon was used, but some experts have suggested that ricin is the most likely candidate. Oleg Kalugin, who confirmed the assassination, described that the KGB had developed a gel with a toxin that was to be smeared on the victim. It was likely that this happened to Solzhenitsyn.

Assassination in a London Street

It was not only the Soviet Union that used bioweapons against unwanted individuals. In 1978, Darjavna Sigurnost (DS) – the secret policc in the Soviet satellite state Bulgaria – carried out a deadly assassination of the Bulgarian defector Georgi Markov. This assassination has received considerable attention.[40, 86]

Markov was a well-known author and dissident in communist Bulgaria. He defected in 1969 and lived in London from 1971, where he worked as a commentator for Radio Free Europe and on the BBC's Bulgarian broadcasts. One day in September 1978, while waiting for a bus at Waterloo Bridge in central London, he suddenly felt a sting on the back of his right thigh. At the same time, he saw a man picking up an umbrella from the street, saying 'sorry'. The man attempted to hide his face and immediately disappeared in a taxi. Afterwards, Markov had a painful injection mark on the back of his thigh and some blood on his trousers.

That evening, Markov developed a fever, nausea, vomiting, abdominal pain, and diarrhoea. The next morning, he was hospitalised but he continued to worsen and died four days later. The autopsy initially revealed no cause of death, however, tissue samples from the injection site examined at Porton Down showed a small metal capsule in the tissue. It had a diameter of only 1.52mm and had a small cavity with two openings that had been sealed with a type of wax. The capsule was made of 90% platinum and 10% iridium – metals that cause minimal tissue irritation. The capsule likely contained ricin, which leaked into the tissue after the wax sealing the openings melted at body temperature. The disease progression in Markov aligns well with ricin poisoning studied at Porton Down in trials with pigs. The ricin capsule was apparently shot into Markov's thigh with a specially made umbrella.

Markov had feared assassinations from the Bulgarian secret police for an extended period. The suspicion after the successful assassination naturally targeted Bulgaria, which however dismissed this as James Bond-inspired fabrication.

Later, former KGB General Kalugin, who also confirmed the assassination of Solzhenitsyn, reported that the Bulgarians had asked the KGB for help in eliminating Markov in early 1978. The KGB evaluated several proposals but ultimately chose the method that led to success in London. According to Kalugin, the Bulgarians had first tested it on a political prisoner. Even the umbrella used was the work of the KGB.

All Bulgarian documents regarding the Markov case were deleted, and the head of the secret police later committed suicide. The agent who executed the assassination has since been likely identified as an Italian with Danish citizenship who had been recruited by the Bulgarians.

The murder of Georgi Markov attracted considerable attention and has inspired books, plays, and TV series. The British government's reaction to this political murder – carried out in the heart of London in broad daylight – was remarkably muted. This was likely partly due to the lack of conclusive evidence of Bulgaria's role in the case.

Murder Attempt at the Arc de Triomphe

The Markov assassination was not Bulgaria's debut on the international stage. Ten days earlier, their secret police had attempted to eliminate another Bulgarian defector, Vladimir Kostov, in Paris.[86]

Kostov was a major in the Bulgarian secret police and operated in Paris under the guise of being a journalist before he defected in 1977. He then participated in German radio and BBC broadcasts to Bulgaria.

On 26 August 1977, Kostov was attacked outside the metro station at the Arc de Triomphe in Paris. Suddenly, he felt a sharp pain in his back, while he heard a loud bang that sounded like a gunshot. His wife discovered an injection mark with some bleeding on his back, and a corresponding hole in his jacket and shirt. Kostov experienced fever over the next few hours but recovered after a couple of days.

When Kostov later heard about the murder of Markov, he underwent an X-ray examination, which revealed a small metal particle lodged in the subcutaneous tissue on his back. The particle was removed and examined at Porton Down. It turned out to be identical to the capsule found on Markov, which likely contained ricin. Kostov had probably been fortunate because the amount of ricin in the capsule had been close to a lethal dose.

Also in this case, there was no solid evidence that the Bulgarian intelligence service was behind it, although the suspicion was strong.

It has been claimed that the same method used against Markov and Kostov has been employed in several other political murders, but information here is uncertain.

An Assassination Attempt in Congo

Assassination attempts on unwanted individuals were not limited to one side during the Cold War. In 1975, the American Congress established a committee led by Senator Frank Church to investigate alleged illegal actions carried out by the CIA, FBI, and National Security Agency (NSA). They sought among other things to investigate the possible use of bioweapons. A long series of shocking facts were uncovered, revealing that the CIA had collaborated for many decades with bioweapon researchers at Fort Detrick on weapons that the CIA wished to use for assassinations of foreign leaders. A number of such assassinations were planned but were mostly carried out in a surprisingly amateurish manner – and they were as far as we know all unsuccessful.[52]

The Congolese politician and liberation leader Patrice Lumumba was one of the CIA's targets. Lumumba was the first head of government of the Republic of the Congo after liberation from Belgium in 1960. After just a few months, he was ousted by the president, and General Mobutu Sese Seko orchestrated a coup.

When Lumumba came to power, he was seen as a dangerous man by the Americans and the Belgians, partly because he had approached the Soviet Union seeking help in combating the uprising in Katanga province.

Possibly due to the fear that the new republic would give the Soviets a foothold in Africa, the CIA planned to kill Lumumba with botulinum toxin. CIA agents arrived in Congo with the toxin but were unable to infiltrate Lumumba's security network. A few months later, the deposed Lumumba was delivered to the rebels in Katanga, where he was executed by shooting under the leadership of a Belgian officer – and presumably with the blessing of the USA.[86]

Castro – the CIA's Favourite Target

CIA's favourite target among foreign leaders was Fidel Castro – Cuba's foremost political leader for half a century. In 1959, the dictator Fulgencio Batista was overthrown during the so-called Cuban Revolution, and Castro established a communist regime that became a thorn in the side of the USA. Nine presidents attempted to overthrow Castro and in this effort, the CIA was a central actor. According to Cuban intelligence sources, the CIA laid 638 plans to assassinate Castro, of which 167 were reportedly implemented. These figures are likely exaggerated. However, the Church Committee found at least eight CIA plans to murder Castro, some so fanciful they could have been taken out of an action film.[52, 86]

They aimed, among other things, to capitalise on the fact that Castro was an enthusiastic scuba diver. The plan was to outfit him with a diving suit laced with a deadly mushroom toxin while the breathing apparatus was contaminated with tuberculosis bacteria. The plan failed because Castro – possibly by mix-up – received an untreated diving suit. Another plan was to supply him with cigars infused with botulinum toxin. Other methods of assassination were also planned involving poison, explosives, and firearms. The CIA partially collaborated with the American mafia on these plans.[107, 108]

All of these assassination attempts failed, and Castro could rightly say: 'If surviving assassination attempts were an Olympic event, I would have taken gold.' He died peacefully in Havana in 2016.

Pablo Neruda's Unexpected Death

Solzhenitsyn is not the only Nobel Prize-winning author who may have been sought eliminated by state actors. In February 2023, it became known that new forensic investigations of the remains of the Chilean poet Pablo Neruda had detected genetic material from *Clostridium botulinum* – the bacterium that produces botulinum toxin. Does this mean that Neruda was done away with using this toxin?

Pablo Neruda died 12 days after Chilean President Salvador Allende was overthrown by General Augusto Pinochet. The official cause of death was prostate cancer, which he evidently had. However, both his family and many friends have been convinced that he was murdered by the Pinochet regime. Neruda was a prominent member of the Communist Party and a friend of Allende. He had reacted strongly to the takeover of power in Chile and had planned to leave the country to go into exile. On 23 September 1973, he died at a clinic where he had been admitted the same day. The day after, he had planned to leave the country. According to his driver, he received an injection of unknown nature a few hours before he died.

Shortly after Neruda's death, the Communist Party of Chile declared that he had been murdered by the regime. In 2015, Chile's now democratic government also stated that it was not unlikely that this was a murder. An international panel of experts was appointed to investigate Neruda's remains for a wide range of poisons without yielding any positive findings. In 2017, another international group was appointed to look for signs of bioweapons. The result was the discovery of genetic material from *Clostridium botulinum*.

Neruda's family has interpreted these findings as conclusive evidence that he was murdered. The Pinochet regime is also believed to have used botulinum toxin to kill several political prisoners in Chile. The famous Neruda would undoubtedly have become an uncomfortable critic of the Pinochet regime in exile.

However, researchers are more cautious in their conclusions. Although the findings suggest that botulinum bacteria were present in Neruda's body at the time of his death and likely did not arise later, that is not evidence that they harmed the poet. If there had been an intent to kill him with botulinum toxin, it is very likely that this would have been injected directly – and then the bacterium would not have been present in the body. He may have unintentionally ingested the detected bacteria with some food at some point. Additionally, not all variants of *Clostridium botulinum* produce the toxin.

Thus, we do not know for certain if Neruda was murdered – and probably we will never receive a definitive answer.

Bioweapons and Non-Political Crime

Politically motivated biocrime committed by states has received the most publicity. However, it is a fact that there are far more cases where bioweapons have been utilised by non-state actors with 'ordinary' motives, such as jealousy, greed, and revenge. The first cases emerged as early as the early 1900s.

It is not surprising that a significant portion of such biocrime has been committed by doctors or others with a specific connection to the healthcare system. These groups, naturally, have greater knowledge of microbes and toxins and often have easier access to them than the general population.

Let's look at some examples of such biocrime.

Deadly Food and Drink

In a highly publicised court case in the USA in 1996, Dr Debora Green was convicted of attempting to kill her husband, heart specialist Michael Farrar, three times by giving him ricin in his food. Additionally, she was convicted of killing two of her three children by setting their house on fire.[86]

After each ricin attack, Farrar was hospitalised. It took some time before the doctors became aware of the cause of his symptoms, which led to serious complications such as brain abscesses and endocarditis (inflammation of the heart valves) that required surgical treatment.

Green's motive for the assassination attempts on her husband was clearly that he wanted a divorce. The marriage had been very tumultuous for a long time, and she also had significant psychiatric problems, including alcohol abuse.

Green refused to disclose how she carried out the ricin attacks on her husband. She received a lengthy prison sentence and can be released in 2035 at the earliest.

In 1938, Japanese doctor Tei-Sabro Takahashi was sentenced to death for causing 17 people to get typhoid fever. Three of them died.[86]

In three of Takahashi's bioweapon attacks, the targets were other doctors whom he viewed as competitors. He aimed to destroy their practices and delivered cakes laced with typhoid bacteria to their offices. The doctors themselves avoided infection, but several people in their surroundings were affected. On another occasion, Takahashi attempted to infect yet another colleague whom he strongly disliked.

Finally, he tried to kill his wife with typhoid bacteria, but she survived. He attempted to kill her in the hospital by injecting her with morphine, but this was prevented. He was exposed by one of the doctors he had tried to infect when that doctor had the cake Takahashi had brought analysed.

Takahashi himself stated that he chose *Salmonella typhi*, the typhoid bacterium, as a weapon because he believed it would be perceived as a natural infection. At that time, typhoid fever was a rather common infection in Japan.

One of the most extensive examples of biocrime through foodstuffs is the repeated use of typhoid and dysentery bacteria by Japanese doctor Mitsuru Suzuki from 1962 to 1964 to harm people in his surroundings.[86]

Suzuki held a degree in bacteriology. He obtained the bacteria he used from a laboratory where he worked for a period and also secured bacteria from infected patients. By various methods, he managed to infect at least 120 individuals, of whom four died. Many of the victims were colleagues at the university hospital where he worked – both doctors and nurses – whom he infected with the bacteria in cakes or bananas, as well as through contaminated drinks. On one occasion, he injected dysentery bacteria into individuals he had already infected with typhoid bacteria, doing so under the pretence that it was treatment. Ordinary patients were also infected.

He did not limit himself to using bioweapons against people at the workplace. Relatives and friends also did not escape. During a visit to his uncle's family, he inflicted typhoid fever upon eight family members. The same happened to his brother and sister-in-law.

Suzuki's motives for these bioweapon attacks appear to be multiple. He explained that his attacks on colleagues were revenge for what he perceived as discrimination in his work situation. The attack on relatives was also revenge, allegedly because they would not support him financially in his efforts to complete his doctoral thesis concerning typhoid fever. He also admitted that some of the cases were because he wanted to test a theory he had about infection.

The typhoid bacterium has also been popular with biocriminals outside of Japan. In 1921, insurance agent Henri Girard in Paris was charged with infecting a number of people with bacteria and toxins. He had his wife, mistress, and two others as accomplices.[86]

Girard was an insurance agent with a rather special business idea: he persuaded clients to take out life insurance where Girard or his accomplices would benefit in the event of death. He then did his best to ensure that the death occurred.

He had studied bacteriology before his career as an insurance agent and chose the typhoid bacterium and a fungal toxin as suitable weapons. He had obtained the typhoid bacterium on the open market, but he also had a small private laboratory in his mistress's house. In three cases where Girard managed to inflict typhoid fever on his clients and their family members, they survived to his great disappointment. In one of the cases, he succeeded through the injection of an unknown toxin.

Disappointed by the uncertain efficacy of the typhoid bacteria, Girard chose to use a fungal toxin against his next client – a widow – whom he managed to persuade to take out four insurance policies. She died three hours after Girard served her an aperitif laced with the toxin. But this time, one of the insurance companies sensed something was wrong, leading to Girard's operation being revealed.

Girard died before the high-profile trial started, but all his accomplices were convicted. His wife received a life sentence, and his mistress 20 years.

Plague as a Murder Weapon

Yersinia pestis – the plague bacterium – has also been used as a weapon by ordinary criminals. In 1935, Doctor Taranath Bhatacharya and Benoyendra Chandra Pandey were sentenced to death by hanging in Calcutta (Kolkata) in India for infecting Benoyendra's younger half-brother with the plague bacterium. The sentence was later commuted to life imprisonment.

The two half-brothers Benoyendra and Amarendra disagreed about the distribution of their father's inheritance, and Benoyendra decided to kill his brother. Together with his friend, Dr Taranath, he chose to use the plague bacterium. Taranath managed to obtain *Yersinia pestis* from a laboratory in Bombay and ensured that the bacteria were truly virulent, meaning they could cause disease, by testing on rats.

As Amarendra was about to take the train from Calcutta, a stranger – hired by Benoyendra – suddenly bumped into him. At the same time, he felt a sting in his arm. It was only after three days that he sought medical attention, where they discovered a needle mark on his arm. Then he developed fever and pain in the armpit on the same side as the sting. He gradually grew sicker and died seven days after the sting at the train station.

Amarendra was cremated, making an autopsy impossible. However, a blood sample taken before he died confirmed the presence of *Yersinia pestis*.

It was quickly clarified that the deceased's half-brother was a friend of the bacteriology-trained doctor Taranath and that he had obtained the plague bacteria. Both were subsequently charged and convicted of the murder.

Murder with Diphtheria Toxin

Bacterial toxins have also been used as murder weapons. In a high-profile trial in St Petersburg in 1911, Patrick O'Brien de Lacy and Doctor Vladimir Pantchenko were convicted of killing Pantchenko's brother-in-law with diphtheria toxin.[86] This toxin, produced by the bacterium *Corynebacterium diphtheriae*, is crucial for the bacterium's ability to cause life-threatening disease.

De Lacy was a Polish nobleman with significant debts due to unfortunate financial speculation. He had married the daughter of a very wealthy general and his mother-in-law was also very rich. He decided to solve his financial problems by killing his in-laws and brother-in-law, allowing his wife to inherit the entire estate.

To carry out these plans, he contacted Vladimir Pantchenko – a doctor with a rather dubious background. They agreed on the price for each of the planned murders. Initially, de Lacy wanted them to use cholera bacteria as a weapon, since cholera was not uncommon in St Petersburg. Pantchenko was sceptical about this and instead suggested diphtheria toxin. This toxin causes severe complications from the heart, kidneys, and nerves. Pantchenko obtained both diphtheria toxin and cholera toxin.

The pair chose to start with de Lacy's brother-in-law. He died after Pantchenko seized the opportunity to inject the toxin, which he claimed was a treatment for cholera. The death appeared suspicious, and de Lacy's father-in-law demanded an autopsy. This led to a full investigation and the arrest of de Lacy and Pantchenko. De Lacy was sentenced to life imprisonment while Pantchenko received 15 years.

HIV Virus – A Popular Weapon

After the AIDS epidemic broke out in the 1980s and the HIV virus was discovered, this virus gained popularity as a potential murder weapon, although the virus's effect can take a long time to manifest.[110] While HIV as a weapon was indeed more dangerous before 1996, when breakthroughs in treatment occurred, infection remains serious. Unsurprisingly, it is doctors and other healthcare personnel who dominate this statistic as perpetrators since HIV blood and syringes are more readily available to these groups than to others. The usual murder motives are also present here.

In 1992, Brian T Stewart allegedly injected HIV blood into his 11-month-old son while the boy was hospitalised, according to prosecutors in Missouri. The motive was likely that Stewart believed he was not the boy's father and therefore did not wish to pay child support to the mother. He worked at a hospital where his job was to take blood samples, giving him access to blood from HIV patients.[86]

The son underwent progressively worsening health issues without doctors understanding the cause. It was not until the son was 5 years old and severely ill with a confirmed HIV infection that Stewart was suspected and investigated. Staff at the hospital where the father had injected HIV blood into the son reported suspicious behaviour at the time. Stewart's colleagues also mentioned that he sometimes threatened to inject contaminated blood into people he was in conflict with.

Stewart was brought to trial and convicted based on strong circumstantial evidence to life imprisonment. The judge expressed regret that he could not impose a harsher sentence.[111] The son survived thanks to the new HIV medications introduced in 1996.

In the same year that Stewart was convicted, gastroenterologist Richard J Schmidt was convicted in Lafayette, Louisiana, for infecting Janice Trahan – his mistress of ten years – with HIV.[86]

Schmidt's motive was that Trahan wished to end their relationship, something he was unwilling to accept. Under the guise of administering a vitamin B12 injection, which he used to give her, he injected HIV blood against her will and left her. After a month, she developed symptoms of HIV infection. Prosecutors were able to demonstrate that her HIV virus was of the same type as in one of Schmidt's patients. Schmidt was sentenced to 50 years in prison.

The cases of biocrime mentioned are just a small selection of the many murders and attempted murders committed with bioweapons. Many other such cases have likely gone undetected. An important aspect of the use of microbes and toxins, both in terrorism and biocrime, is that the victims' symptoms can be perceived as a result of naturally occurring infections – and the connection is therefore never clarified.

Chapter 11

Armed Against the Rider of Plague? How to Respond to Bioweapon Attacks?

In the ancient world, little could be done against outbreaks of deadly epidemics. It is not without reason that the plague on its pale horse is one of the four horsemen of the apocalypse mentioned in the Book of Revelation. These dreadful riders were given power 'to kill with sword, famine, pestilence, and by the beasts of the earth'. The alliance between war and plague may also allude to the use of plague as a weapon.

Biological weapons have, as we have seen, a long and dramatic history. They have been used both in war and in terrorism, as well as in ordinary forms of crime. It is entirely unrealistic to assume that bioweapons will not be used again in such contexts, although there may be some disagreement regarding the magnitude of the risk. It is therefore necessary to have well-thought-out plans for how society should respond to a biological attack. The fundamental principles of bio-defence are the same in military and civilian contexts. Nonetheless, we face the greatest challenges in planning the defence against bioterrorism in civil society, where such an attack can have catastrophic consequences.[112] We will therefore focus on this.

The First Hours and Days

In the case of terrorist attacks using conventional as well as chemical and nuclear weapons, the effects occur immediately and are entirely apparent. However, with terrorist attacks using microbes, there will be a certain amount of time before the effects become noticeable.[113,114] This latency period is due to the fact that infections have a certain incubation time – the time between infection and the outbreak of the disease – which varies from microbe to microbe. This is, of course, an advantage for the terrorist, who may have plenty of time to escape.

However, this latency period can also be a potential advantage for those who must respond to the threat of terrorism. This applies in cases where the microbe used as a weapon spreads easily from person to person and can gradually trigger a widespread epidemic. If one can identify the microbe quickly upon the first cases of illness, a number of measures can be put in place to contain and eventually stop the terror-triggered epidemic, which will take some time to develop.

Rapid identification of the microbial weapon is therefore crucial for a successful defence that limits the damage.[115] In practice, this depends on healthcare personnel who first encounter the terror-induced cases of illness suspecting an outbreak of infection that could either be natural or a result of bioterror. Particularly important here are health personnel in intensive care and infectious disease departments, but also other parts of the primary healthcare system that will encounter the first cases of illness early on. An effective response from front-line personnel must also involve collecting necessary samples from patients so that the microbe can be identified as quickly as possible.[114,115]

In certain cases, specific characteristics of the disease in the affected patients may raise suspicion of the microbe responsible and indicate that it is not a natural outbreak of an infection but rather a case of bioterror.[113,114] This is relevant if the disease profile suggests an infection that is not found commonly in the region. An example of this is the anthrax attack in the United States in 2001, where a knowledgeable infectious disease physician in Florida, as mentioned previously, suspected that one of the first cases was due to the rare disease of pulmonary anthrax.

In most cases, however, identification of the microbe will depend on laboratory diagnostics. It is therefore crucial that diagnostics occur rapidly and efficiently in the microbiological laboratories that receive the first samples. In recent years, there have been significant advancements in the development of rapid methods for such diagnostics, where the genomic structure of microbes can be analysed quickly through so-called gene sequencing and thus identify the microbe. Such methods should be available.[114]

For several years, work has also been done to develop special equipment that can detect certain microbes of particular interest in terrorism in a very short time.[112,115] Such instruments can be deployed in specific public places to provide early warning of microbial threats. So far, there have been no successes in American cities where this has been attempted, partly due to false alarms from the instruments.[116] Portable instruments of this type may be more useful in the event of a possible bioterror attack, as this can provide early identification of microbes or toxins. Until now, such instruments have primarily been of military interest, but they will also be useful for civil bio-defence. The drawback of such technological quick-fix solutions, however, is that they will only detect one or a few relevant microbes. Nevertheless, new and improved types of instruments are under development.[117]

Bioterror or Natural Outbreak?

Once the microbe has been identified and characterised, the results may in some cases lead to suspicions that it is a case of bioterror rather than a natural epidemic outbreak. Several factors regarding the microbe and the forms of infection it has triggered can be important in this assessment.

As mentioned previously, the microbe might be unusual in the area it has affected.[115] For example, if an outbreak of the plague bacterium were detected in Oslo – where plague has not occurred for several hundred years – this would undoubtedly provide the definitive answer. It would also be suspicious if the microbe affected a large number of people simultaneously over a broad area, which could occur in an effective aerosol attack. If the microbe is more contagious or pathogenic than the natural 'versions' of it, or if the microbe affects unusual age groups, this could suggest that it has been altered in a laboratory and is thus a product of bioterror. Investigations of the microbe's genome, i.e., its genetic material, could also raise suspicions that it has been 'tampered with' to make it even more dangerous. Efforts are under way to develop rapid methods to detect such suspect characteristics in microbial genomes, although it may not always be possible to identify man-made changes in microbes.[118]

Unfortunately, the recognition that an epidemic outbreak is not natural but human induced often comes slowly for healthcare personnel. One example of this is the bioterror attack initiated by the Rajneesh cult in Oregon, where epidemiologists and health authorities long – despite prolonged and thorough investigations – dismissed the possibility of bioterror.

Even if healthcare personnel may have some suspicion that an epidemic outbreak could be due to terrorism, it is often the police who, after thorough investigations using conventional methods, are able to conclude that biological terror is likely. Among other things, discoveries of bioweapons laboratories, witness interrogations, identification of suspicious perpetrators, and information gathered from social media will be useful here. In some cases, the perpetrators will eventually claim responsibility but this is more common in terrorist attacks using conventional weapons.

While the first warnings of a potential bioattack will often come from healthcare personnel, routine surveillance systems tracking disease occurrences in the population may also provide early signals that something serious is happening on the infectious front. For instance, in Norway, a number of infectious diseases are reportable, and weekly reports must be sent regarding infection occurrences from various institutions. Some countries like the USA have attempted to develop such monitoring systems further, routinely gathering and correlating data on a wide range of infection-related health data – symptoms and signs of illness, medical visits, hospital admissions, laboratory findings – from emergency departments, outpatient clinics, and laboratories. This form of monitoring, known as syndromic surveillance, is hoped to provide early warnings of bioterror attacks. However, it is resource intensive and has not fully met expectations.[115] A similar monitoring system was also launched in Norway during the Covid-19 pandemic.

Technological solutions and various types of monitoring systems can undoubtedly be useful but early warning of bioterror attacks will often rely on vigilant and well-prepared healthcare personnel. Education about bioterror should be provided in medical school, along with later courses for the most relevant personnel groups.

Major Challenges for Healthcare

An effectively executed terrorist attack using bioweapons could lead to a sudden increase in the number of seriously ill patients requiring acute hospital admission. In many cases, this will involve infections that require special isolation and care to prevent further transmission. It is therefore necessary to have plans in place for how the healthcare system can quickly increase its capacity to meet these challenges. Some infectious diseases require highly specialised isolation rooms, but there are very few of these. Consequently, it may be necessary to devise new, improvised solutions regarding isolation and care settings and the use of buildings when existing hospital capacity is insufficient. Plans must exist for this.[115]

Furthermore, it is essential to have sufficiently trained healthcare personnel for the care and isolation of infection patients and for intensive care of the most severely ill, such as patients with infection-induced respiratory failure. Healthcare personnel will also be particularly vulnerable to infection while working with terrorism victims. It is therefore crucial that they have effective protective equipment against infection.

In Norway, serious capacity issues in these areas were revealed when the Covid-19 pandemic was at its peak. A bioterror attack would very quickly present significantly greater problems than this pandemic. Therefore, there is a considerable risk that the healthcare system would become overwhelmed in the event of a severe bioterror attack.

Vaccines and Antimicrobial Drugs – Often in Short Supply

What options do we have for treating the infectious problems of the numerous seriously ill victims in the event of a bioterror attack? Such people will require hospitalisation, and many will require treatment in intensive care units, including the use of ventilators in cases of respiratory failure.

As for medical measures directed specifically against the microbe or toxin used in the bioattack, vaccines and medications are primarily relevant. However, significant challenges may arise here.[114, 115]

We have effective antibiotics for most of the bacteria currently considered potential weapons, such as anthrax, plague, and tularaemia. It is crucial that treatment begins rapidly for infections such as pulmonary anthrax and plague. If so, the majority of patients can be saved, hence adequate stockpiles of such antibiotics must be available to ensure preparedness.

In the context of terrorism, we cannot overlook the possibility that bacteria may be genetically modified to become antibiotic resistant. The Soviet Union has already developed several resistant strains of anthrax bacteria in its Biopreparat programme.

The challenges posed by viruses are considerably more significant than those arising from bacterial infections. In modern infectious medicine, we still lack effective, well-tested medications for many viruses, including those that may be of interest to terrorists.

Variola, or smallpox, is one of the viruses still feared as a weapon. Although this virus has been eradicated globally, it is stored in two laboratories in the world where control over it is maintained. However, with modern biotechnological methods, it is possible to synthesize the smallpox virus, thus making it available to terrorists. There is one medication that is promising in the treatment of the smallpox virus – tecovirimat – but it has not been tested in large clinical trials.[115] In most countries, including Norway, tecovirimat is not available for widespread use during a terrorist attack involving smallpox, which would undoubtedly lead to a high number of illnesses. Since smallpox vaccination was discontinued more than 40 years ago, the global population currently lacks immunity against this virus.

The drug ribavirin has some effect against several haemorrhagic viruses that could potentially be used as weapons, such as the Ebola virus – but extensive clinical experience is lacking. Many other medications with potential efficacy against various viruses are still in the laboratory phase.

This lack of effective medications against viruses means that today, victims infected with viruses in a terror attack may primarily be offered general treatment measures to maintain vital bodily functions until the patient's immune system potentially gains control over the infection.

In addition to medications, vaccines can also play a role in combating a bioattack. Generally, vaccines are used to protect individuals against infectious diseases before exposure occurs, by increasing the immune response against the microbe, making it easier for the body to combat any subsequent infections.

Were we able to vaccinate the population in advance against those microbes considered the most relevant as weapons, it would reduce the impact of an attack significantly. The grim reality, however, is that today we only have well-tested vaccines against a handful of the most relevant microbes in terms of terrorism. Furthermore, the few vaccines available are not accessible to the public and are not intended for mass vaccination. Among these is the anthrax vaccine, which has so far only been used in military contexts. As we have seen, several thousand US soldiers received the anthrax vaccine before the second Gulf War. Many also received a vaccine against botulinum toxin. Vaccines exist against the smallpox virus, which is indeed a potential weapon. A relatively new smallpox vaccine has fewer side effects than the old vaccine used until the late 1980s, but it is currently not available for extensive use.[115]

In an acute situation following a bioterror attack, a vaccine could also be beneficial.[114, 115] For several infectious diseases, vaccination shortly after exposure may prevent disease or at least reduce the severity of the illness. This applies, for example, to smallpox, which could lead to a catastrophe in a bioattack if the infection spreads uncontrollably. In such a situation, vaccination of all exposed individuals would commence as soon as possible, along with their contacts through what is known as 'ring vaccination', thereby limiting the spread of the infection. The vaccine must be administered within the first two to three days after exposure to be effective. Quick action is thus necessary. Rapid identification of the pathogen is therefore crucial.

It is of great importance for the defence against terror to develop additional vaccines for relevant microbes, although we are unlikely ever to be able to obtain vaccines against all microbes and toxins that may be employed by terrorists in the future. Modern biotechnology could also enable terrorists to develop new variants of 'old' microbes that are resistant to existing vaccines. This was pursued by the Soviets during the Cold War, as we have seen. Furthermore, terrorists might employ entirely new microbes as weapons, particularly among the many newly discovered microbes causing 'emerging infectious diseases'.

During the Covid-19 pandemic, we saw that biotechnological advancements led to a new type of vaccine: mRNA vaccines. Such vaccines can be developed significantly faster than conventional vaccines, but probably not fast enough to stop the initial damage from a bioterror attack involving a new microbe.

Both conventional and new vaccines are based on what is called immunological memory, meaning the immune system's ability to 'remember' contact with a microbe after initial exposure, and then react with a strong immune response during any future contact. However, it will take some time to establish such immunological memory.

Some have therefore suggested developing methods to stimulate what is called the innate immune response – components of the immune system that initiate the fight against a microbe immediately, without relying on prior contact with the microbe. Such treatment would have a rapid effect and could potentially be useful in a bioterror attack.[119] However, it has yet to be implemented in the treatment of infections in humans.

Organisation and Management

Combating an acute bioterror attack thus requires efforts from a number of groups and institutions – health personnel, police, hospitals, and health authorities.[115] A robust response to bioterror is therefore dependent on effective organisation and coordination of efforts, with stable and efficient leadership from the authorities. This also includes ongoing and fact-based information to address the fear and uncertainty that will arise in such a situation. Such information is also crucial to ensure public understanding of any drastic measures that must be implemented, such as quarantine and isolation measures, school closures, and restrictions on the movement of the population.[115] During the Covid-19 pandemic, we saw how sensitive many of these measures could be, as well as the necessity of ongoing and considered information from the authorities.

Essential prerequisites for the effective combatting of a bioterror attack are that clear, detailed plans exist, and that necessary stockpiles of medication, vaccines, and protective equipment against infection are in place. In many of these areas, we noted deficiencies during the recent pandemic.

In the USA, in June 2001, it was decided to test preparedness against bioterrorism through a comprehensive exercise, Dark Winter. This involved a simulated attack with the smallpox virus. The exercise revealed serious problems in several areas.[120] The USA has since conducted four additional exercises simulating bioterror attacks – the last two in 2018 and 2019 when a new coronavirus was the supposed weapon. All exercises exposed problems in the preparedness for bioattacks. Despite these valuable learning points, the USA's response to the Covid-19 pandemic revealed many of the same deficiencies that were revealed in the aforementioned exercises.

In Norway, similar extensive exercises have not been conducted. We would probably face the same issues in the event of a potential bioterror attack as were observed in the American exercises. At the outbreak of the

Covid-19 pandemic in 2019, it became apparent that Norway's pandemic plans were simply outdated.

There is no doubt that a large-scale bioweapon attack could have catastrophic consequences for society. Therefore, maximum effort should be invested in preventing such attacks. The international biological weapons convention (BWC) applies only to states and is not relevant for biological terror, where defence relies on individual nation-states. The USA and some other states have therefore enacted several laws directly applicable to bioterrorism. Primarily, however, it will be the monitoring and investigation by each state's intelligence services – in Norway, the PST (Police Security Service) – that will be central in preventing bioterror. Input from vigilant citizens regarding suspicious activities will be of great value.

Chapter 12

Gaia as Victim – When Bioweapons Attack Animals, Plants, and Ecosystems

In Greek mythology, Gaia is the primordial goddess from whom all gods and all living things originate. She is Mother Earth herself – the personification of the Earth. In the 1970s, James Lovelock and Lynn Margulis developed the so-called Gaia hypothesis, which views the Earth – Gaia – as a living superorganism that must be preserved and protected. Among humanity's many activities that can harm Gaia are biological weapons.

When bioweapons are discussed, the primary focus is typically on their use against humans. However, this type of weapon can also be directed against animals and plants. The state programmes developed in the twentieth century included such combat agents, even if they did not receive the same priority as bioweapons aimed at humans. Nonetheless, they are also covered by the Biological Weapons Convention (BWC).

Let us briefly look again at the bioweapons that were aimed at animals and plants that were considered significant in state programmes.

State Actors

Germany had an extensive bioweapons programme during World War I that was solely directed toward animals important for the enemy's warfare, particularly horses and mules.[24] During the war, horses were especially crucial for transporting supplies and equipment for the fighting armies. Some biological attacks were also aimed at cavalry horses. German operations in Finnmark and Finland may also have targeted reindeer.

The Germans primarily used the bacteria *Bacillus anthracis*, which causes anthrax, and *Burkholderia mallei*, which causes glanders. The attacks were conducted both in hostile countries like the USA and France, as well as in neutral countries that supplied animals to Germany's opponents.

Germany's biological warfare scarcely had any significant impact on the course of the war.

France had a rather modest bioweapons programme during World War I. This programme, too, was aimed at enemy animals, primarily Germany.

As we have seen, the UK invested significantly in the development of bioweapons during World War II. While a variety of microorganisms were assessed, the focus eventually narrowed to anthrax. By the end of the war, only the substantial stockpile of 'anthrax cakes' that had been built up was ready for use. The intention was to spread these via aircraft over the main cattle areas in Germany to decimate significant portions of the livestock population and thus impact the country's food supply. Some humans would likely also be affected by anthrax, but humans were not the primary target. The anthrax cakes were never used.[34]

In the extensive bioweapons programme that the USA built up in the years following the war – until President Nixon halted it in 1969 – bioweapons aimed at various plants essential for global food supply were also included. The Americans focused on fungal species that could harm plants. Two of these could damage wheat and rye, and a third could affect rice plants. This suggests that military operations were also planned in Asian countries.[34]

The white minority regime in Rhodesia also used bioweapons in its struggle against liberation forces. There is considerable evidence that the extensive anthrax epidemic in Rhodesia in 1978 – the largest in recent history – was due to the use of anthrax as a weapon.[54] The intention in this case would be to target meat production in the tribal areas where the liberation forces had bases.

The Soviet Union's Biopreparat programme is one of the most extensive the world has seen. It also included bioweapons aimed at animals. In addition to the bacteria *Burkholderia mallei* and *Burkholderia pseudomallei*, which cause disease in horses, several viruses targeting animals were part of the Soviet arsenal: African swine fever virus, which causes illness in pigs and wild boar; Capripox virus, which affects goats and sheep; Rinderpest virus, which affects cattle and other ruminants; and the Hand-Foot-and-Mouth Disease (HFMD virus, which affects cattle, sheep, goats, and pigs.[47]

In a military conflict, the Soviet Union was thus well prepared to strike at food production in both wealthy and developing countries. We do not know for certain whether these bioweapons were ever used.

It also turned out that Iraq's bioweapons programme, which created much uncertainty during the two Gulf Wars, included a fungal toxin that could harm wheat plants. Neither this nor other items from Iraq's bioweapons arsenal were utilised.[34]

It is probably no coincidence that many states have included bioweapons targeting plants and animals in their programmes. If used effectively, there is little doubt that such weapons can inflict significant damage on the enemy's food supplies, which can give the attacker great strategic advantages.

Agroterrorism

Terrorist attacks using bioweapons against plants or animals are termed agroterrorism. In principle, agroterrorists may have similar motives and goals as terrorists using conventional weapons. When purely economic motivations are decisive, it can be referred to as more ordinary biocriminality – but the boundaries can be fluid.

It will often be relatively easy for agroterrorists to obtain relevant microbes, which are frequently available in nature. Moreover, many of the microbes of interest will be harmless to humans, making it less risky for terrorists to work with them. Furthermore, the relevant microbes can often be used directly without special treatment in laboratories – weaponisation – in laboratories. It is also easier to deploy bioweapons against plants and animals than to carry out attacks on humans.[121, 122, 123]

The risk of detection is also lower with agroterror than in terrorist actions targeting humans. It is often very difficult to distinguish between natural and human-made outbreaks of infections when plants and animals are afflicted by infectious diseases.

Since such terrorist attacks are not aimed at humans, the threshold for agroterrorism is likely to be lower, and it may not be met with the same moral condemnation as other forms of terror. All of this makes agroterrorism potentially attractive to terrorists.

Which Countries Are Vulnerable?

All countries can be severely affected by terrorist attacks on agriculture for various reasons.

The way food production is conducted industrially in most Western countries makes it highly vulnerable to attacks. In the USA, for example, dairy farming is conducted in facilities housing between 1,500 to 10,000 animals, which live in extremely close quarters with high potential for spreading infectious diseases. Chicken production operates under similarly industrial conditions in massive facilities. Animals' resistance to infectious diseases has decreased in recent years due to a range of factors, such as antibiotic use and hormone treatments, and vaccination against animal diseases is often not performed. Extensive transport of animals could lead to the effective spread of infectious diseases introduced by terrorists.[124]

In many developing countries, the economy is much more based on agriculture than in Western countries. In countries like Afghanistan and Yemen, about 75% of the population is engaged in agriculture. This also makes these countries very vulnerable to terrorist attacks on plants and animals.

So far, only a handful of such attacks have occurred. However, it would not be surprising if agroterrorism becomes more common in the years to come. It is known that terrorist organisations have already considered these possibilities. After the USA invaded Afghanistan following 11 September 2001, hundreds of Al-Qaeda documents translated into Arabic were found that referenced aspects of American agriculture. This suggests that the terrorist organisation considered this a potential target for attack.[122]

The few examples of agroterror known to date have been relatively modest in scope, and they have been executed ineffectively and unprofessionally.

The Mau Mau Uprising and Bioterror

An early example of bioterror aimed at livestock comes from the Mau Mau uprising against the white colonial regime in Kenya in the 1950s. The rebellious movement, primarily based on the Kikuyu tribe, called itself the Kenya Land and Freedom Army (KLFA). The uprising ignited in 1952, and the British declared a state of emergency. In the ensuing years, both sides engaged in battles marked by great ruthlessness and brutality until the uprising was effectively quelled by 1956. From the Mau Mau side, attacks were directed at British farmers, involving the murder of Europeans and mutilation of livestock.

In 1952, 33 oxen at a mission station were attacked with the toxin from the plant *Synadenium grantii* (African milk bush). This plant was well-known to the Kikuyu. Eight of the animals died. The perpetrators had evidently rubbed the milky sap containing the toxin into cuts on the animals' skin. Trials with the toxin that the British tested on guinea pigs showed that the animals exhibited the same symptoms as the oxen that had been attacked.[86, 125]

This was clearly part of the Mau Mau insurgents' systematic assaults on the colonists' livestock. Arsenic poisoning was also used for this purpose. The use of bioweapons is not documented in other cases during the uprising, but several livestock deaths of unclear cause might have been examples of this.

Meat-Eating Maggots as Bioweapons

For centuries, cattle ranching in Central America and in the southern parts of the USA has been plagued by the parasitic fly *Cochliomyia hominivorax* – known as the 'screwworm'. This fly lays 200–500 eggs in wounds on warm-blooded animals, including humans. The larvae develop from the eggs within the tissue, causing inflammation as they burrow inwards. The animal becomes sick and often dies.

In 1985, an active programme was launched to eradicate the screwworm in Mexico. Mexican agricultural workers involved in this feared for their jobs in the programme, probably since this brought extra income, which would be lost if the programme succeeded and the parasite was eradicated. To prevent this, they began to spread the fly larvae among cattle.[125]

This occurred just a few miles south of the border with the USA, which feared that it could also affect American cattle ranching. Investigations into the screwworm outbreak led the US Department of Agriculture to conclude that it was a bioterror attack carried out by Mexican agricultural workers.

Farmers Against Rabbits

In August 1997, many dead rabbits were discovered on a farm in New Zealand. It turned out that the cause of death was rabbit haemorrhagic disease (RHD). This infection is caused by a deadly and highly contagious virus – *rabbit calicivirus* – that primarily affects rabbits.

It was revealed that the virus had been illegally introduced from Australia and used to infect the rabbit population in New Zealand. The background to this was that rabbit numbers had increased enormously in New Zealand since the late 1800s and had become a significant economic and environmental problem for farmers. They received no public support for measures to combat the rabbit plague, and the costs were vast.[86]

In Australia, which had battled its rabbit population for years, the RHD virus had been used as a means to reduce it. Farmers in New Zealand had sought government permission to do the same but were denied, even after appealing. This denial created great dissatisfaction and was criticised not only by farmers' leaders but also by the Prime Minister and the Minister of Agriculture.

The farmers then chose to take matters into their own hands. It remains unknown who originally introduced the RHD virus into New Zealand. However, it is clear that a significant number of farmers collaborated to systematically spread the virus in the country's rabbit population.[86]

Tissue from rabbits that had died from the infection was ground up and mixed with carrots, grains, and other rabbit feed that was distributed on farms where the rabbits were prevalent. Dead rabbit carcasses were also laid out.

Initially, the authorities attempted to combat the spread of the viral infection. However, this effort failed, and it gradually had to be accepted that the virus had come to stay in New Zealand. It had become endemic.

No one was prosecuted or punished for the action, and none of the active farmers have expressed any regret for their actions. Regardless, the illegal introduction of the RHD virus into New Zealand is perceived as an example of bioterrorism.

Sewage as Bioweapon

The conflict between Israeli settlers and Palestinians in the West Bank is a significant international problem that has yet to be resolved. The Israeli settlements are considered illegal under international law but continue to increase. In 2000, Palestinian news sources reported that settlers were actively directing large amounts of sewage water onto Palestinian agricultural areas to drive Palestinians away. The sewage water has considerable destructive effects on the areas, causing dead olive trees and problems for livestock on the Palestinian side.[125]

If the spreading of sewage water was indeed used as a means in the conflict between Israelis and Palestinians, it can be seen as a primitive example of bioterror.

Threats of Bioterrorism

There are certain examples of serious threats of agroterrorism. In 2011, a South African man was arrested and later convicted for threatening to spread the HFMD virus that causes foot-and-mouth disease in the USA and the UK if he was not paid $4 million. This was allegedly an act of revenge for these two countries' failure to prevent white farmers from losing their land in Zimbabwe.[125]

As we will discuss further, infection with the HFMD virus can impose substantial economic damage. The threat was therefore taken seriously. However, investigations revealed that it was an empty threat, as the responsible individual did not have access to the virus.

What Can Be Achieved with Agroterrorism?

Over the years, there have been claims of the use of bioweapons against plants and animals. Nevertheless, very few such attacks are documented convincingly. However, there is good reason to believe that terrorists will resort to bioweapons in the years to come. The goals of many terrorist groups will not only be to take human lives but also to inflict significant economic damage on 'hostile' countries. For example, Al-Qaeda's former leader, Bin Laden, was convinced that attacks aimed at the American economy would prevent the country's international military activity and therefore should be central to the organisation's strategy toward the USA. There is reason to believe he considered agroterrorism an interesting option.

Regarding attacks on animals, there is a substantial number of microbes that are of interest to terrorists. In addition to the bacteria that featured in the weapons programmes of the USA and the Soviet Union, there are 12–15 viruses that stand out. These include several viruses that are highly contagious and have a high mortality rate.[124, 126]

One relevant example is the previously mentioned, highly contagious HFMD virus. This can spread through both direct contact and via objects, and it can also be transmitted via aerosols over several kilometres. Several natural outbreaks of this virus have frighteningly illustrated the significant damage an epidemic can cause. During the outbreak in the UK in 2001, 6.5 million animals – cattle, sheep, and pigs – had to be slaughtered. The costs associated with this are estimated at £8 billion. During another outbreak in South Korea from 2010 to 2011, over three 3 million animals were culled, with economic impacts estimated at $2.6 billion.[126]

Among the many other viruses that terrorists could utilise are Newcastle disease virus and variants of avian influenza virus, both of which could cause great issues for chicken production.[124, 126] The damaging effects of the avian influenza virus have been demonstrated recently in a considerable outbreak among seabirds in Norway.

The possibility of agroterrorism directed at plants has received even less attention than attacks on animals, but it is certainly a significant threat. In nature, there are several thousand microbes – primarily fungi and bacteria – that can infect and damage plants. Estimates suggest that such plant diseases cost the USA around $33 billion. The use of such microbial weapons would therefore be highly appealing to terrorists wishing to strike at a country's economy.

The list of relevant fungi and bacterial species that could be used by terrorists in plant attacks is very long. It is not difficult to obtain the microbes, as they exist in nature.

Such a bioattack would not only have economic consequences but could also lead to serious psychological and political repercussions if the population loses confidence in food security. If countries that usually export large quantities of grain are affected, this could also have major implications for food situations in countries reliant on imports. In connection with the war in Ukraine, we are already observing global issues that could arise if exports are impacted in one of the world's most significant grain-exporting countries.[124, 127]

Bioweapons, Biodiversity, and Ecocide

The term Anthropocene has recently been introduced to describe the period in which we currently find ourselves. The very designation 'Anthropocene' is derived from the Greek word for human – anthropos – and suggests that this period is characterised by the extensive impact of human activities on the biological and physical conditions of the planet. Important examples include climate change due to the emission of greenhouse gases, deforestation, and pollution of soil, seas, and air. This has had dire ecological consequences, including loss of biodiversity, referred to in many discussions as the threat of so-called ecocide – a total collapse of entire ecosystems.

The use of bioweapons is also one of the many human activities that can cause serious harm to nature. This is particularly true for the use of biological weapons against animals and plants, which can have extensive ecological repercussions.[121, 123] Such weapons would not only affect livestock populations and agricultural plants; in many cases, the microbes could also spread to wild animals and plants with devastating effect – potentially leading to the extinction of species without resistance to the microbe. History has many examples of microbes spreading over large areas with destructive impacts on both large and small animal groups. Especially rare, previously endangered species are vulnerable. Thus, the use of bioweapons could lead to reduced biodiversity and, in the worst-case scenario, ecocide.

Genetic homogeneity is often significant among animals and plants of great nutritional importance in agriculture, but it varies across different regions. Genetically selective bioweapons could thus have devastating effects.

Many biological weapons designed to target humans could also spread to animals, both domesticated and wild. Anthrax, tularaemia, and plague – which form the basis for 'classic' biological weapons – are zoonoses, just like many other infections that affect Homo sapiens.[13] They originate in the animal world and could return there when used as bioweapons.

How to Prevent Agroterrorism?

It is unlikely that 100% effectiveness in preventing well-planned bioterror attacks can be achieved. The first condition for effective prevention and combatting is to consider the possibility and, as with other bioterrorism, to develop plans for it. A concerted effort from many personnel groups and institutions will be necessary, and this work must be coordinated. Veterinarians and agricultural experts with insight into plant and animal diseases will play a central role here.

There are relatively few treatment methods for plant and animal infections. American military researchers associated with the Department of Defense have therefore attempted to explore new avenues, particularly in the light of possible terrorist attacks against the country's grain production. In a project mentioned previously, they aimed to utilise specific insects that feed on plant sap. These insects are used as carriers of viruses that can alter the genetic make-up of plants, making them resistant to microbes that could be employed by bioterrorists. As we have seen, this project has been met with considerable criticism from other researchers, who fear that this method could also be used to develop bioweapons against plants. This presents an ongoing dilemma in research primarily aimed at developing defences against bioweapons.[73]

Regardless, the combating of bioweapons is necessary not only to protect ourselves – the species Homo sapiens – but is also crucial to prevent serious damage to Gaia, Mother Earth.

Chapter 13

Frankenstein Returns – An Ethical Dilemma in Modern Research

In 1818, the English author Mary Wollstonecraft Shelley published the novel *Frankenstein; Or The Modern Prometheus.* It has become a classic and has been adapted into film numerous times. In this book, which has elements of both science fiction and horror, Shelley tells the story of the gifted scientist Victor Frankenstein, who attempts to create an improved human type in his laboratory. His experiments go awry, leading to the creation of a terrifying yet intelligent monster. Shelley does not detail Frankenstein's experiments, but it is likely that he used parts from animals in constructing the creature. Regardless, the consequences are dreadful. The monster becomes a killing machine, with Frankenstein's intended bride among its victims. Consequently, Frankenstein attempts to destroy the monster he has created but fails. In the final scene of the book, the monster drifts away on an ice floe in the Arctic darkness, while Frankenstein is left helpless.

Shelley's novel has often been interpreted as a warning against humanity going too far in its 'manipulation' of nature. This objection is frequently raised against modern natural sciences and medical research, which largely aim to alter nature for human benefit. Can such research lead, in certain cases, to unforeseen catastrophic consequences? Can it not increase the threat from biological weapons?

Research as a Double-Edged Sword

In discussions about arms control and disarmament, the term 'dual use' is used to describe technology that can serve both civilian, 'useful' purposes but can also have military applications.[128, 129] Similarly, new research findings can become a double-edged sword. This is a central issue in biological research, which primarily aims to gain new insights into fundamental life processes.

This knowledge can lead to improved welfare and health for humans, animals, and plants but can also provide the foundation for the development of biological weapons.

Risky Research

Following the anthrax attacks in the USA in 2001, concerns about bioterrorism increased. This led the American National Academy of Sciences to establish a committee to assess the risks associated with 'dual use' in biological research. The committee published its report in 2003 under the title *Biotechnology in an Age of Terrorism*. This Fink Report, named after committee chair Gerald Fink, described a number of experiments it deemed particularly risky forms of 'dual use' research.[128, 129] The Fink committee concluded that such experiments must be assessed carefully and monitored and recommended the establishment of a new advisory body for researchers wishing to conduct this type of research. This led the American president to create the National Science Advisory Board on Biosecurity (NSABB) in 2004.

Based largely on the Fink committee's conclusions, NSABB described seven categories of risky biological research identified as DURC – 'dual use research of concern'. This is research that can:

enhance the pathogenicity (virulence) of a microbe or toxin,
reduce the effectiveness of vaccination against a microbe,
confer resistance in a microbe or toxin against useful medications or make it more difficult to detect them,
increase the infectivity or potential spread of microbes or toxins as biological weapons,
increase the number of animal species that can be affected by a microbe or toxin,
increase susceptibility to a microbe or toxin, or
create entirely new microbes or toxins or 'revive' an extinct microbe.

As we have seen, several of the experiments on this list were relevant to the development of biological weapons in the former Soviet biological weapons programme. Such experiments have also not infrequently been conducted later in other areas of research, examples of which we will later examine.

Awareness of this was undoubtedly part of the driving force behind NSABB's warning about DURC.

Avian Influenza Virus Increases Concerns

Concerns surrounding DURC were initially confined to a smaller circle in the research community, and there was limited publicity about this issue. This changed in 2011 when researchers from two leading influenza laboratories, Ron Fouchier from the Erasmus Medical Center in the Netherlands and Yoshihiro Kawaoka from the University of Wisconsin in the USA, reported astonishing results from their experiments with the avian influenza virus H5N1.[129, 130]

This virus, discovered a few years earlier, primarily infects wild birds and occasionally domestic birds such as chickens and ducks. Rarely, it can also infect humans living in close proximity to infected birds. In humans, this is a very dangerous infection with a lethality rate of around 60%. Fortunately, the virus is very poorly transmissible from person to person. Therefore, larger epidemics have not occurred so far, just isolated cases among those close to the initially infected individuals.

However, due to the high lethality associated with H5N1 infection in humans, there has been significant concern that the virus could change its properties through mutations and become much more easily transmissible from person to person. This could lead to an unimaginable pandemic catastrophe worldwide since really effective medications against the virus do not exist.

What researchers Fouchier and Kawaoka achieved – by somewhat different methods – was to genetically modify the H5N1 virus such that it could now spread easily through aerosol and droplet transmission among ferrets, a common laboratory animal used in influenza research.[130] The fact that this dangerous virus could be altered in the laboratory to become transmissible among mammals raised alarms among researchers. They feared that increased transmissibility could also apply to humans. Would not this altered virus be a terrible bioweapon if it fell into the hands of terrorists? And would not a catastrophic pandemic be the result if such a virus accidentally leaked out of the laboratory?

The heated discussion that followed these H5N1 experiments led a number of scientists, including Fouchier and Kawaoka, to agree in early 2012 to pause all such research for some months while they discussed the issues. NSABB also urged Fouchier and Kawaoka to withhold certain information when publishing their research results to prevent others from replicating the experiments. This led to strong protests about censorship, and ultimately the articles were published, with all details, in 2013.[130]

Risky Research Paused

In 2013, the same types of experiments with both the H5N1 virus and other dangerous avian influenza viruses resumed, which raised considerable concerns once again.[130] However, what really had implications and led to the intervention of American authorities was a series of incidents involving dangerous microbes in several laboratories, including Kawaoka's in Wisconsin. Authorities decided that from 2014, for a three-year period, no government funding would be granted for experiments of the type that had triggered the debate – experiments involving influenza viruses and, in addition, the coronaviruses SARS and MERS, which have high mortality rates but limited transmissibility in humans.[130]

The controversial experiments that had been in the spotlight were referred to in the public debate as gain-of-function research (GOFR). This stemmed from laboratory manipulation that added new functions to the virus. In reality, GOFR is a subgroup of what NSABB called DURC. However, GOFR may not be a fortunate designation because it is too imprecise. This can make it difficult to identify the most dangerous forms of DURC, while at the same time creating problems for more benign research.

The period of prohibiting presumed risky research was used to develop new regulations for assessing risky GOFR experiments. At the end of 2017, the ban on GOFR was lifted, and a new approval system for such experiments was put in place. Now, a special committee within the Department of Health would carry out this assessment at the request of one of the federal institutions, the National Institutes of Health (NIH) and the Centers for Disease Control (CDC). This particular committee is often referred to by its acronym P3CO (Potential Pandemic Pathogen Care and Oversight).

In the new committee guidelines, GOFR was significantly narrowed down to refer only to research that could reasonably lead to a modified microbe likely to be highly contagious and could give rise to a pandemic. Additionally, it was required that the microbe would likely be highly virulent and could lead to significant morbidity and/or mortality in humans. Moreover, the guidelines applied only to 15 microbes and toxins on the government's list of so-called select agents.

These guidelines for controlling GOFR have been subject to strong criticism from several researchers who argue that far more effective monitoring and oversight of such research is necessary. Critics have put forward several arguments.[132]

Firstly, they criticise the notable lack of transparency in the process of assessing potentially risky experiments that could be prohibited. The names of committee members are strictly confidential, and the same goes for the discussions within the committee, including the projects under consideration.

Critics have also pointed out that only three projects have been approved so far, while numerous projects that, according to the guidelines, should have been evaluated have slipped through the process. This includes certain Covid-19-related projects carried out or planned in Wuhan, China, which involved collaboration between American and Chinese researchers, partly with government funding from the USA.

But the most central argument against the current system is the risk that especially risky GOFR experiments may trigger serious pandemics – either through laboratory accidents that lead to microbes 'leaking' into the environment, or through manipulated microbes with increased pandemic potential giving rise to bioweapons.[133, 134] These arguments have gained strength amidst the heated debate regarding the origin of the Covid-19 virus, where one hypothesis suggests that the virus is a result of GOFR conducted at the Institute of Virology in Wuhan. Critics argue that any potential benefits of GOFR are clearly outweighed by the risks, and that other less risky forms of research could yield similar results.[135]

Recently, critics have organised themselves into a separate organisation called Biosafety Now! that advocates for stricter oversight of GOFR both nationally and internationally.[136] It has been argued against Biosafety Now! that the organisation lacks researchers working actively in the relevant areas.

On the other side of this ongoing debate, a number of scientists have pushed back strongly against calls for enhanced regulation of GOFR – including research with clear risks.[137, 138] The primary argument of these scientists, which significantly includes virologists who themselves work in the relevant research fields, is that GOFR can generate new knowledge of great significance in the fight against microbes, including prevention of new pandemics. This includes the development of new vaccines and medications against infectious diseases, which is needed urgently. Such invaluable knowledge cannot always be obtained through other research methods, these researchers argue.

Furthermore, opponents of stronger research regulation emphasize that GOFR is an imprecise term and difficult to apply practically to specific research projects. New functions developed in microbes as a result of laboratory experiments do not necessarily lead to increased pandemic risks, and it is often impossible to predict the consequences of such changes. Each individual research project must therefore be assessed individually with respect to benefit and risk; rigid, centrally based systems for evaluating GOFR may not be targeted enough. Such systems may, on the contrary, hinder important and necessary GOFR and thereby have significant detrimental effects on microbiological research and the fight against infectious diseases. The alternative these critics seek is that researchers themselves and individual research institutions should bear the main responsibility for assessing risks associated with GOFR.

The heated debate surrounding GOFR has persisted for several years and shows no signs of diminishing. Both the discussion about the origin of the Covid-19 virus and certain recently conducted research projects perceived as risky have led NSABB to work on new guidelines for assessing and regulating GOFR. A new draft was discussed by a number of experts at a meeting in January 2023.[139] In this draft, NSABB proposed, among other things, greater transparency in assessments than the current system allows and that it should not be limited to the 15 microbes and toxins previously in focus but should extend to many more viruses, bacteria, and fungi. Furthermore, it was suggested that research conducted in other countries that is funded by the USA should be subjected to more thorough oversight, undoubtedly inspired by the debate surrounding the origin of the Covid-19 virus.

Not surprisingly, there was no consensus regarding the NSABB proposal among the experts at this meeting. It was decided that NSABB would make changes to the draft and return with a new proposal. It is entirely uncertain when this will happen and what a potential new draft will entail.[139]

Laboratory Leaks of Dangerous Microbes

A central argument in the discussion about the risks associated with GOFR, which is also clearly relevant to the issue of biological weapons, is the risk of laboratory accidents that can cause dangerous microbes to 'leak' out of laboratories. The fact that such accidents can, and do, happen quite frequently is under-communicated in public debate.[130, 133]

It is evident that work in laboratories with infectious microbes must be carried out with a certain level of caution. However, the stringency of safety regulations depends on how dangerous the microbe is – its virulence and transmissibility. There is general international consensus about what safety level is required for various microbes. Laboratories are classified, as mentioned earlier, into four levels depending on the extent of the safety level (biosafety level or BSL).[140] In BSL-1 laboratories, the lowest safety level, the work is with the least harmful microbes where only simple precautions against infection are required. At the next level – BSL-2 – work is conducted with microbes that pose a moderate risk of infection and serious illness, e.g., hepatitis viruses, malaria parasites, and some bacteria. Here, the rules are somewhat stricter.

At the highest levels – BSL-3 and BSL-4 – work is conducted with microbes that can cause serious, potentially life-threatening diseases, especially through airborne transmission. At these levels, there are specific requirements for ventilation conditions. BSL-3 is considered necessary for work involving several viruses, including the coronaviruses SARS-CoV-1 and SARS-CoV-2 and yellow fever virus, and several bacteria, including plague, tuberculosis, and tularaemia bacteria.

In BSL-4 laboratories – which are depicted frequently in the press and in films – personnel wear 'space suits' with their own oxygen supply. In these laboratories, work is conducted on microbes that transmit easily via the respiratory tract and often cause life-threatening infections for which there are no medications or vaccines. For instance, Ebola and Marburg viruses, smallpox virus, and several other exotic viruses fall within this category.

The higher the safety level, the more strenuous it is to work in the laboratory – and the more expensive the operations.[130]

On paper, the detailed safety requirements for work with microbes – even the very dangerous ones – are reassuring. In addition to technological safety measures, thorough training of personnel in the laboratories is required. It may be surprising to many that accidents and errors that pose risks of dangerous microbes escaping – and, in the worst case, triggering pandemics – still occur.

American data about such accidents are likely the most reliable, although they are also incomplete. Between 2009 and 2015, a total of 749 laboratory accidents involving dangerous microbes were reported in government laboratories. What is striking is that at least 80% of these incidents were due to human error.[141] These could involve accidental injuries, such as personnel getting pricked by infected needles or being bitten by an infected laboratory animal. However, the accidents may also stem from negligence, such as deviating from safety protocols. This may partly be due to the significant strain of working in BSL-3 and BSL-4 laboratories. The consequences of such incidents could lead to dangerous microbes 'leaking' out of laboratories. This could happen in several ways. Infected laboratory personnel may take microbes home, potentially infecting others. Microbes that have not been destroyed per regulations can also find their way out of laboratories.

In December 2023, results were released from a review of all published laboratory accidents worldwide from 2020 to 2021.[142] A total of 94 such incidents were identified, where 309 individuals were infected with a total of 51 different microbes. In eight cases, the infections were fatal. In 16 cases, the infection spread outside the particular laboratory. In one of these cases, where *Brucella* bacteria 'leaked out', 10,528 individuals in the area became infected. Like the American surveys, human error was the cause in the majority of cases. The reported incidents likely represent only a fraction of actual events due to incomplete reporting in the vast majority of countries.

While human error is the most frequent cause of laboratory accidents, there can also be failures in the complicated technology necessary in high-security laboratories, such as specific ventilation systems. Laboratories require careful maintenance. The previously mentioned anthrax disaster in Sverdlovsk in 1979 illustrated the consequences of a failure in ventilation systems, but in that case, human error was also involved in the maintenance.[28]

It is therefore beyond doubt that high-security laboratories where dangerous microbes are worked with pose a significant risk to public safety. This pertains to the environments surrounding the laboratories but also to the possibility of far more extensive epidemics – in the worst case, pandemics. It is therefore a matter of considerable concern that a number of new BSL-4 laboratories have been built in many countries in recent years. In 2021, there were reportedly 59 such laboratories spread across 23 countries. The number of BSL-3 laboratories is unknown, but it is estimated that there are more than 1,700 in the USA alone. There are no binding international regulations for the operation and maintenance of these laboratories. Such regulations are also lacking in a majority of countries that possess such facilities. Many are located in countries with unstable regimes and turbulent societal conditions.[143, 144]

What is actually going on in the many high-security laboratories around the world is not known. The possibility that biological weapons development is occurring in some of them cannot be ruled out. Furthermore, the risk of a 'leak' of dangerous microbes is present if GOFR is ongoing in the laboratories. Microbes can also go astray through regular theft committed by corrupt laboratory personnel or by unauthorised groups gaining control of the laboratories. The risk naturally increases if GOFR is being conducted. Through such laboratories, terrorist groups could acquire biological weapons.

The Covid-19 Virus and the 'Lab Leak' Debate

The gain-of-function debate has been ongoing for decades, but until recently it was primarily confined to the research community. With the outbreak of the Covid-19 pandemic in 2019, this debate intensified, engaging the public and media, particularly American politicians. The background for this renewed interest in GOFR is the uncertainty surrounding the origins of the Covid-19 virus. There are two main theories being discussed in this context.

Many researchers believe that the Covid-19 pandemic is a classic zoonosis, meaning that the virus originated from animals and then infected humans. We have numerous examples of this.[13] As mentioned previously, this applies to the majority of newly discovered emerging infectious diseases.

Several viruses associated with zoonoses originate from bats. In folklore and popular literature, these silent, nocturnal creatures have been associated with horror. They often appear in such tales alongside demons, witches, and wizards. The existence of blood-sucking bats in some parts of the world also evokes associations with the Dracula legend and vampires. While we can likely exonerate bats from connections to supernatural powers, it is a fact that many species of bats are a natural reservoir for a wide array of viruses, including many coronaviruses.

Proponents of the zoonosis theory believe that the Covid-19 virus stemmed originally from a specific type of bat and that it then infected humans via another animal species – an 'intermediate host'. This is thought to have occurred through the now-famous wet market in the city of Wuhan, where a wide variety of both domestic and wild animal species have been sold to satisfy Chinese culinary interests.[145]

The alternative theory – the 'lab leak' theory – which also has many supporters, posits that the virus leaked from the Wuhan Institute of Virology (WIV) in Wuhan, China.[146, 147] This institute is China's flagship in virus research and has conducted extensive research on coronaviruses from bats for years, including viruses that are quite closely related to the current pandemic virus.[147] The central scientist here is Shi Zhengli, who has been nicknamed Bat Woman. There is evidence suggesting that this research at WIV also involved risky GOFR, but this is a contentious issue. Proponents of the lab leak theory assert that the Covid-19 pandemic is a clear example of the disasters that can ensue as a result of GOFR, thus illustrating the necessity for stricter oversight of such research.[133, 148]

There is no clear evidence supporting either of the two hypotheses regarding the origin of the Covid-19 virus. The main argument for the zoonosis theory is that this is a well-documented mechanism for new infectious diseases. Furthermore, it is claimed that there is no proof that coronaviruses present in the WIV laboratory were related closely enough to the Covid-19 virus to be a likely precursor. Claims that the structure of this virus shows signs of having been 'manipulated' in the laboratory are also not accepted.[149]

Supporters of the lab leak theory base their argument on a number of factors they believe point clearly to the virus originating from the WIV.

They note that there have been prior leak incidents of the original SARS virus (SARS-CoV-1) from several Chinese laboratories, which in one case led to nine infections and one death.[146, 147] They also emphasize that work with coronaviruses at WIV has evidently not been conducted according to international regulations that dictate that such viruses require at least a BSL-3 laboratory, even though WIV has both BSL-3 and BSL-4 laboratories.[147, 150] Notably, there have been striking inaccuracies and secrecy in the laboratory's description of the viruses they have worked with in recent years. Supporters of the lab leak theory also stress that it has not been demonstrated that any animal at the wet market in Wuhan was infected with the Covid-19 virus. The virus has not been found in thousands of animals that could have been intermediate hosts for the virus's transmission from bats to humans.[147]

The investigation into the potential role of WIV in the outbreak of the Covid-19 pandemic has been extremely challenging because China has not wanted to cooperate fully. Chinese authorities have refused to provide extensive insight into the laboratory's operations and other conditions in Wuhan that are relevant to understanding the pandemic's origins. On the contrary, Chinese officials have indignantly dismissed the possibility that the virus could have originated from their laboratory. An international commission of scientists appointed by the World Health Organisation (WHO) to investigate the origins of the Covid-19 virus was not granted access to all relevant material at WIV and has therefore been unable to draw any final conclusions.[13, 146, 147]

The debate regarding the origin of the pandemic virus has become very bitter and polarised. This is partly due to a conflict within the research community and also the specific domestic political conditions in the USA.

In its coronavirus research, WIV has engaged in extensive collaboration with scientists in several countries. For several years, American virus researchers have collaborated with Chinese virologists in Wuhan on projects that many perceive as risky forms of GOFR. Several of these collaborative projects have been funded with significant amounts of money from the American government, including through the National Institutes of Health (NIH). An important recipient of these funds has been the American non-governmental organisation EcoHealth Alliance, which aims to prevent zoonoses, i.e., infectious diseases that spread from the animal kingdom to humans.[147, 150] It has gradually become clear that this organisation bases a portion of its research on risky GOFR, including collaborative projects carried out at WIV.

The lab leak theory thus implies that American scientists and the money of American taxpayers may have contributed to the Covid-19 pandemic – something many would perceive as a scandal. This is likely an important reason why a group of involved researchers, who initially considered the lab leak theory plausible, publicly denied it. In articles in leading scientific journals, they labelled it a mere conspiracy theory – evidently to quell a debate about the virus's origins and research collaboration with WIV. Several of the researchers behind these submissions had participated in GOFR in Wuhan, including the head of EcoHealth Alliance, Peter Daszak.[147, 150]

This strategy initially had a considerable impact on the public debate regarding the virus's origins. One of the reasons for this may have been the fear of losing access to research funds, since central figures behind the campaign controlled important funding sources, including the NIH and the Wellcome Institute in the UK, which together are the largest contributors to research in the world. However, the strategy only succeeded for a few months before it was exposed as an attempt to suppress debate through means unbefitting scientific discourse. A science journalist[147] and a number of prominent scientists[151, 152] pushed back strongly, asserting that the lab leak theory was a legitimate alternative to the zoonosis hypothesis and not an absurd conspiracy theory. Partly at the urging of the authorities, big tech – the major information technology companies such as Facebook, Google, and Twitter – engaged in a veritable censorship of all posts promoting the lab leak theory.[153, 154]

It is hardly coincidental that researchers who dismiss the lab leak theory are generally opponents of extensive regulation of GOFR, while proponents of this theory advocate for regulation.

The dramatic nature of the discussion around the origin of the Covid-19 pandemic has been intensified significantly by the fact that the discussion within the research community has also become entangled in the highly polarised political debate between the Democratic and Republican parties in the USA.

The fact that the leak theory initially received significant publicity is partly attributed to the fact that it had proponents among some top politicians in President Donald Trump's administration. This undoubtedly led many Trump opponents in the press to also dismiss the theory. The situation now is that questions about both GOFR regulation and the origin of the Covid-19 virus have largely been politicised.[155] On the Republican side, both the desire for stronger control of GOFR and the belief in the lab leak theory are strong. The Senate has held a number of hearings initiated by Republican senators where several researchers from both sides of the debate have been called to provide testimony.

A central figure in the debate regarding both the origin of the Covid-19 virus and GOFR is the powerful Anthony Fauci, who has been the director of the National Institute of Allergy and Infectious Diseases (NIAID) for nearly four decades. Fauci, one of America's leading immunologists and infectious disease experts, has been a strong proponent of GOFR research – including its more risky forms.[156] He has also played a central role in attempting to debunk the lab leak theory – even calling for censorship by big tech.[147, 148]

After the CIA established its own committee to assess the lab leak theory, it has been alleged that Fauci held secret meetings with this intelligence agency.[157] The majority of the CIA committee was initially clearly in favour of the lab leak theory but reportedly changed their views after receiving bribes. After this, the CIA's official position was that it was impossible to take a stand in this debate.

Fauci has thus been called to several hearings in the Senate where he has had to answer questions from aggressive Republicans regarding both the GOFR issue and NIAID's support for research at WIV. The temperature during these hearings has been notably high.[148] That he is often seen as primarily responsible for the many unpopular measures during the Covid-19 pandemic, including the aggressive use of Covid-19 vaccines, does not enhance his popularity among Republicans, as even public health measures have become a polarised political issue in the USA. Stringent measures are associated with the Democrats among certain segments of the population.

The controversial nature of Fauci's position in the USA is illustrated by the fact that he is almost canonised among his supporters as America's saviour during the Covid-19 pandemic. He has received numerous awards for his efforts, and idyllic children's books have been written about him. On the other hand, his opponents claim he shares responsibility for the pandemic through the lab leak from WIV and is therefore to blame for the loss of millions of lives. He has been accused of perjury – lying under oath – repeatedly during hearings in Congress. Fauci and his family have received death threats and require bodyguards.

The crucial debate over DURC and GOFR in research should have been conducted in a calm and rational manner. It is regrettable that it has been contaminated by political and personal motives. There is indeed a risk that this could undermine public trust in both science and health authorities in the long term. Recent surveys in the USA have shown that this trust has been declining since the onset of the Covid-19 pandemic.[158]

'Biohacking' and 'Invisible Laboratories' – A New Problem

In addition to the unresolved issues regarding effective monitoring of activities in professional biological laboratories, a new problem has emerged in recent years: the establishment of simple, laboratories where amateurs can carry out their own research – so-called do-it-yourself (DIY) research or biohacking.[159] Although these amateur researchers typically lack formal education and training in biological research, they can acquire knowledge, among other sources, from the internet about many new biotechnological research methods.

Biohacking has become an international movement with at least 90 laboratories in various countries, hosting conferences and websites dedicated to the practice. Those who utilise these laboratories work on a variety of biological issues. Some of the projects involve working with bacteria, which are altered using genetic engineering techniques according to the project's objectives.

Such biohacking projects may occasionally have practical utility. However, it is also clear that risk factors are associated with these DIY laboratories.[159, 160]

Essentially, these laboratories operate at the simplest safety level – BSL-1. However, work is carried out with altering the functions of microbes, so the potential for gain-of-function issues and the possibility of leaking such microbes is certainly present in these basic laboratories. It is a problem that in most countries, including the USA, which has the most DIY laboratories, there is no established system for monitoring and oversight of them. Authorities have no overview of these laboratories – they are effectively invisible.[160]

It is also concerning that such laboratories could be utilised by bioterrorists to produce microbes as biological weapons. Although the production of high-profile bioweapons –'weaponised' for use as aerosols – requires technology that DIY laboratories lack, it is still possible to produce microbes that can be used through more primitive means of dissemination, as seen in previous examples. While the typical use of such laboratories is amateurish, the possibility that potential bioterrorists with biotechnological education could seek out these venues to avoid oversight cannot be ruled out.

The dangers associated with DIY laboratories and biohacking were demonstrated in the autumn of 2023. Government officials in the city of Reedley, California, serendipitously discovered a laboratory where laboratory equipment, test animals, and an array of pathogenic microbes – HIV, dengue, hepatitis viruses, malaria parasites, and Chlamydia bacteria, possibly even Ebola virus – were found.[160, 161] It turned out that the simplest safety measures against the spread of these microbes were not in place. This laboratory was owned by a Chinese citizen without a residence permit and with close ties to Chinese authorities. This has raised concerns and led to calls for monitoring of DIY laboratories, which government authorities have so far shown no interest in.

DIY research may have positive aspects in that it can stimulate interest in biotechnological research among amateurs and may even lead to valuable new discoveries in some cases. However, the aforementioned risks necessitate a form of control. So far, this has been met with resistance from representatives of the DIY movement.

How to Increase Safety in Biological Research?

The situation today is that the issues surrounding DURC and GOFR are not even resolved in the USA, where this debate has primarily been conducted. In the rest of the world, awareness of these issues is much lower, and effective control systems for risky biological research are lacking in most countries. It is therefore necessary to address these problems on an international basis, which is not simple.

The issues surrounding DURC and GOFR are very complex. Nevertheless, it is clear that modern biological research, which has experienced explosive development in recent decades, is associated with significant dangers related to both the issue of biological weapons and the risk of severe laboratory accidents. For the sake of public safety, stricter monitoring of biological research is necessary. How this can be carried out without harming research will pose significant challenges.

There is a reflex among active researchers that political control and regulation is an evil that must be avoided. The adverse effects of such regulation are illustrated, among other things, by the Lysenko tragedy in the Soviet Union. However, our recent research history unfortunately shows that scientists themselves may not necessarily be equipped to assess important ethical issues. Scientific ambitions, fanatical patriotism, and racial ideology, as seen in the history of biological weapons, are examples of motives that can overshadow ethics. The debate surrounding the origin of the Covid-19 virus has also shown that many researchers place less emphasis on concerns for public safety when defending their own interests.

In the years following World War II, all medical research involving patients became subject to ethical scrutiny. It is necessary that ethical research oversight mechanisms are implemented in biotechnological research as well, given the risk factors that such research generally poses today. Research expertise is undoubtedly important in this assessment but it cannot be left solely to the researchers. Also, representatives from public interests must contribute to ensure democratic oversight, in addition to the participation of ethics experts. With a small rewriting of the famous statement by the French politician Georges Clemenceau that war is too important to be left to generals, one could say that research is too important to be left to scientists.

Control mechanisms to prevent risky biological research should be established on a national basis in all research-active states, but they should also be coordinated at the international level. This is extremely complicated to achieve. Such control should also include clear common rules for oversight of high-security laboratories.

Furthermore, it is crucial that awareness of the risk of certain forms of biological research is present among funding authorities when approving research projects, and in scientific journals when evaluating submitted articles.[134, 137]

Chapter 14

On the Battlefield of Propaganda – Bioweapons in Information Warfare

Propaganda refers to the deliberate manipulation of people's feelings and thoughts using powerful means to promote specific views and/or patterns of behaviour. Often, the goal is to spread particular political positions. Propaganda has been known since antiquity, but it gained particular significance throughout the twentieth century with the rise of modern mass media. The development of social media in recent years has further increased the opportunities for effective propaganda.

The history of bioweapons shows that claims and accusations concerning the possession and use of such weapons have been a favoured theme in state propaganda, especially in situations of war. This is due to several factors.

As mentioned previously, after a major or minor outbreak of an epidemic, there will often be doubt as to whether it is natural or man-made, that is, triggered by bioweapons. Here, bioweapons clearly differ from conventional, nuclear, and chemical weapons, where there is usually no doubt that such weapons have been used. Allegations that bioweapons have been employed are often difficult to refute.

Furthermore, the use of bioweapons throughout history has frequently been perceived as immoral and unchivalrous, and this attitude is prevalent even today. In the introduction to the Biological Weapons Convention (BWC), it states regarding biological weapons that 'such use would be morally repugnant to the conscience of mankind and that no effort should be spared to reduce this risk'. Accusations of bioweapon use in propaganda are thus motivated by a desire to provoke global condemnation of the accused party – and perhaps also to justify aggressive countermeasures. There are many examples of this.

The Black Death

An early example of bioweapons use in propaganda can be found during the Black Death in the mid-fourteenth century. In several countries in Europe, Jews were accused of causing the epidemic by poisoning wells. This led to extensive massacres of Jews in many places. Tens of thousands lost their lives. Because of this, a large number of Jews emigrated to Eastern Europe, where they formed communities that persisted until World War II and the Holocaust. In some countries, lepers were also persecuted, as they were suspected of colluding with the Jews.[13]

The Korean War and American Bioweapons – Claims and Facts

The Korean War from 1950 to 1953 was the first major military conflict between the two power blocs during the Cold War. In May 1951, North Korean forces under Kim Il Sung, after an initially successful assault on South Korea, had been pushed back, and ceasefire negotiations had begun. At that time, North Korea claimed to the UN that the USA had spread smallpox in the north during its military operations. In February 1952, North Korea sent an official communication to the UN alleging that the USA had used bioweapons. Shortly thereafter, China fully supported the North Korean claims, which were supposedly confirmed by Chinese scientists who had conducted local investigations where it was claimed that bioweapons had been used. China now launched an intense propaganda offensive against the West based on these accusations.[162]

The USA was accused of having dropped both insects and small rodents that were carriers of the bacteria causing cholera, plague, and anthrax in a series of airstrikes. Allegedly, a number of microbes causing plant and animal diseases were also released. The Soviet Union fully supported these accusations, including at the UN.

China was also able to publish alleged confessions from several American airmen in captivity, who detailed the use of bioweapons. Some of the captives were featured heavily in communist propaganda films and broadcast programmes. The propaganda from China and North Korea also claimed that Americans experimented with bioweapons on prisoners of war.

In their claims regarding the USA's use of biological weapons, the Chinese also reminded that the USA had sheltered Ishii Shiro and his men from Unit 731, who led Japanese biological warfare operations in China. Furthermore, they pointed out that the USA had never ratified the Geneva Protocol, which included bioweapons.

American authorities categorically and indignantly rejected the communists' claims. They claimed that this propaganda itself was a 'weapon of terror', which 'was not only directed at the USA but at the very structure of mankind's civilisation'.

The Americans also believed that the airmen's confessions could not be given any weight, as they had likely been made under severe pressure and threats. Various stories circulated that Western captives in Chinese prisons had been subjected to 'brainwashing' – a term that was becoming popular during this period.[163]

Upon their return to the USA, the American airmen retracted their earlier confessions, stating that they were the result of threats, mistreatment, and torture. However, it has been claimed that even these statements were made under pressure, though this has not been documented. Regardless, the airmen's declarations were forwarded to the UN.

Controversial Investigative Commissions

When the UN, under Norwegian Secretary-General Trygve Lie, proposed in 1952 that WHO or the International Red Cross investigate the accusations against the USA, this was rejected by China. The reasoning was that these organisations were not impartial. However, the Chinese accepted that two commissions – one a legal commission set up by the Soviet-controlled International Association of Democratic Lawyers, and another organised by the also Soviet-controlled World Peace Council at a meeting in Oslo – should investigate the accusations against the USA. In reality, it was China that had initiated this process.[164, 165]

The legal commission made one trip to China and another to South Korea. The commission concluded – based solely on claims from its hosts – that the USA had conducted biological warfare and that it constituted 'genocide and an extremely heinous crime against humanity' that 'surpassed the barbarity of Hitler's Germany and Hirohito's Japan in the last war'.

The second international commission – often referred to as the ISC (International Scientific Commission for the Investigation of the Facts Concerning Bacterial Warfare in Korea and China) – consisted of six members, only one of whom –a Russian KGB general – was a bacteriologist. The most well-known member was Joseph Needham – a prominent English biochemist and sinologist – but also a Marxist. The committee was therefore also referred to as the 'Needham Committee'.[164, 165]

The committee spent two months in Korea and China. It conducted local inspections and interviewed Chinese scientists, the local population, and the American airmen who had confessed in captivity. In its report of over 600 pages, the committee concluded that the USA had used bioweapons as indicated in the original accusations, except that it found no evidence of the spread of smallpox.[162, 165]

The conclusion of the Needham Committee received extensive publicity. In the West, it was met with significant scepticism – even though it was accepted by some with communist sympathies. But Joseph Needham, who particularly advocated for the committee's report, became the subject of massive criticism in England from media, researchers, and authorities, who tried to discredit him as pro-communist and sympathetic to China, which he undoubtedly was.[164, 165]

Needham also had a personal interest in maintaining a good relationship with China since he was in the process of changing his research field from chemistry to the history of Chinese science – a field in which he later became one of the world's leading experts.[165] One might argue that he was naive in trusting unreservedly the assertions of Chinese scientists without considering that they could be under pressure from the Mao regime. Nevertheless, he demonstrated remarkable courage and did not waver in this storm. It has even been said that he had 'the courage of Galileo', referring to the seventeenth-century astronomer's fearless stance in the face of the Inquisition.[165]

However, the critics of the ISC report had many compelling arguments. The committee members were not outstanding researchers and generally lacked substantial expertise in the field of bioweapons. This also applied to Needham himself. During their stay in China and Korea, they relied entirely on the statements of the Chinese researchers. In their local investigations, they were guided carefully by the Chinese, who did not allow them to see the most recent sites of alleged bioweapon use. The interviews with the American airmen in captivity were conducted in fairly amateurish ways in the presence of the Chinese. The alleged American use of bioweapons was also somewhat primitive with some bizarre elements, including the claimed use of disease-carrying insects and small rodents, resembling the earlier biological warfare conducted by the Japanese in China and Manchuria.[166]

Subsequently, information has emerged that further discredits the Needham Committee's conclusions. In 1998, access was obtained to Russian documents that suggest strongly the Soviet Union in 1953 was aware that the accusations of the USA's use of bioweapons were constructed as anti-American propaganda with significant Russian contributions. It appears that the power struggle between Khrushchev and Beria following Stalin's death influenced the Soviet Union's manoeuvres in this case.[162] Subsequent information from China also indicates that the entire affair was a massive propaganda operation against the USA.[167]

Today, the official view – in at least Western countries – is that bioweapons were not used by the USA during the Korean War. By this time, the USA likely did not possess particularly effective bioweapons. The use of insects as carriers of microbes was indeed later considered by Americans, as mentioned earlier. It is possible that the many outbreaks of epidemics that occurred during the Korean War had natural causes related to the warfare. Several of the microbes allegedly used, such as anthrax and plague bacteria, occur naturally in these areas. Once again, we encounter the problem of distinguishing between natural and man-made outbreaks of epidemics, making it difficult to assess claims about the use of bioweapons.

The Olson Tragedy

Nevertheless, some remain convinced that bioweapons were used by Americans in Korea. This is still the official Chinese view but there are also supporters of this perspective in the West. Some of them refer to the tragedy that befell an individual, Frank Olson, who may have had ties to the Korean issue.[168]

Olson was one of the veteran researchers at Fort Detrick and had worked for many years with bioweapons, including anthrax. He was an expert in aerosol technology and had participated in many of the major tests of microbe dissemination on American soil mentioned previously. In 1953, he was also affiliated with the CIA, where he held security clearance. Thus, he gained insight into the CIA's deep secrets, including research programmes testing various substances affecting brain functions on humans in brutal experiments conducted at CIA centres in Germany. There is reason to believe that he began to experience moral dilemmas due to his connection to the CIA's many dubious projects.

In 1953, Olson began showing signs of serious psychological stress. He expressed a desire to leave the CIA, of which his superiors took notice. They evidently began to fear that Olson would reveal secrets from the CIA and Fort Detrick that they believed could harm both the CIA and national security. Olson was called in for discussions with both superiors and a CIA-affiliated doctor. In connection with this, he was in New York, where he stayed at a hotel with another CIA employee. In the middle of the night, he fell from a 15th-floor window, with his death officially declared as a suicide.

The CIA officer who led Olson's treatment was Sidney Gottlieb. He was also responsible for the failed assassination attempt on Patrick Lumumba at that time. Gottlieb had been involved in a long series of the CIA's dubious projects. When he later had his biography published, the book's telling title was *Poisoner-in-Chief: Sidney Gottlieb and the CIA Search for Mind Control.*[169] There is evidence that Gottlieb, during the Korean War, was involved in work related to bioweapons that the CIA may have intended to use there.

Frank Olson's relatives reopened the case after many years. Several circumstances were uncovered that give strong suspicion that it was a murder committed by CIA employees as Olson was perceived as too great a risk. Not only was he familiar with unethical experimental techniques but perhaps also with the American use of bioweapons in Korea, of which his closest family was convinced.

The propaganda war about bioweapon use in Korea took place as the Cold War was reaching its climax. Although the allegations of such use were never substantiated, there is no doubt that they damaged the USA's reputation in the world – both in communist countries and in former colonial nations – where the accusations were given credence. The disagreement over possible bioweapon use during the Korean War illustrates that such accusations can be extremely useful in the propaganda war during military conflicts.

It is also interesting that a central political figure like Henry Kissinger has claimed that China, through its propaganda campaign regarding bioweapons, may have reduced the risk of the USA using nuclear weapons during the Korean War.[164]

The 'Yellow Rain' Affair – Honeybees or Bioweapons

After the end of the Vietnam War in 1975, the Americans withdrew from Southeast Asia. A specific ethnic group, the Hmong people, who lived in the mountains of Laos and North Vietnam, had been allies of the Americans in the struggle against the communists. After the war, they became the target of reprisals from the new communist regimes in Vietnam and Laos, which launched a regular war against them, including airstrikes on Hmong villages.[170]

From 1979, several reports began to emerge about a mysterious, deadly 'yellow rain' from planes during attacks on the Hmong people. It appeared as a sticky precipitation that left small yellow spots on the ground and on the victims. This also occurred in Cambodia, where the Vietnamese fought against Khmer Rouge forces. Representatives from both the Hmong people and Khmer Rouge reported attacks where the mysterious 'yellow rain' led to several hundred deaths. After the Soviet Union invaded Afghanistan in 1979, similar reports emerged from mujahideen fighters fighting against the Russians.[170]

Thousands of Hmong people eventually fled to Thailand, where they were collected in refugee camps, many with injuries supposedly caused by the mysterious 'yellow rain'. Those who were affected described high fever, rashes, sores, breathing difficulties, and bloody vomiting and diarrhoea.

In 1979, the USA sent a commission of military doctors and toxicological experts to Thailand. The commission conducted interviews with Hmong refugees, who provided them samples of the yellow spots on bark, leaves, and stones after the communist airstrikes. Blood and urine samples from suspected victims were also collected. In the USA, chemical analyses of the samples were conducted. In a sample from Cambodia with uncertain background, a toxin – *trichothecene* – was found in one laboratory, which is produced by certain fungal species in nature.[34, 170]

Based on the interview reports from the refugee camps and the discovery of the fungal toxin, US Secretary of State Alexander Haig accused the Soviet government of supplying its communist allies in Southeast Asia with the trichothecene for use as bioweapon. This was denied, but American authorities have never changed their view on this.[170]

The UN also set up an expert committee that interviewed refugees and collected samples from possible victims. The committee's 1982 report concluded somewhat vaguely that there was no evidence of the use of biological or chemical weapons, but could not exclude it either.

In the USA, the highly regarded American molecular biologist and geneticist Matthew Meselson was also drawn into the discussion about 'yellow rain'. He had engaged over several years with issues regarding bioweapons and was sceptical of the USA's claims about the use of such weapons in Laos and Cambodia. In collaboration with various experts, he eventually found evidence suggesting that the explanation for the allegedly deadly 'yellow rain' was very different from the official American version.[170]

Thorough investigations of the yellow spots on leaves and bark, allegedly caused by 'yellow rain' from airstrikes, revealed that they contained pollen that was likely digested by a particular species of large honeybees found in Southeast Asia. Thus, the yellow substance was the bees' faeces. What the Hmong people and Khmer Rouge fighters perceived as 'yellow rain' from enemy planes was actually the result of a large swarm of bees emptying their intestines simultaneously! This could be confirmed visually by Meselson and an insect researcher. A Canadian biologist could later explain this peculiar phenomenon.[171]

It also turned out that the alleged detection of the fungal toxin trichothecene was considerably more uncertain than originally assumed. The methods for detecting this toxin were unreliable at the time. Results from various laboratories showed considerable variation and could yield positive findings in blood and urine samples from people who had never been exposed to 'yellow rain'. Additionally, the toxin-producing fungi were naturally present in Southeast Asia.[34,170]

Thus, the USA claimed that it was the Soviet Union that had supplied the communist regimes with trichothecene. However, based on what we know today about the Soviet biological weapons programme, which was broad, it did not include toxins of this type.

Meselson's explanation sparked considerable debate but was eventually accepted by most. The honeybee hypothesis is undeniably both elegant and convincing but does it explain everything? Is Meselson's hypothesis perhaps a bit too enticing?

In recent years, several previously classified documents have been released from American archives. Based on these documents and a new review of the original facts, some bioweapons experts believe that either biological or chemical weapons in some form may have been used in Southeast Asia.[172]

Was Cuba's Agriculture Attacked by the USA?

The USA, with the CIA as a central actor, was very eager to eliminate Fidel Castro, the undisputed leader of the Cuban revolution, and planned several assassination attempts. But did the Americans also use bioweapons to attack Cuba's agriculture and population, thus undermining the country's economy? Cuban authorities have maintained this stubbornly.

According to the Cubans, the USA has used bioweapons a total of 11 times against both animals and plants of great significance to the country. Let's briefly examine these allegations and the documentation for them.[173]

In 1962, the USA was accused of attacking Cuba's chicken production with the highly contagious Newcastle disease virus, although no documentation was presented from the Cuban side. The cause of this virus outbreak may have been a failed locally produced vaccine that contained live virus.

In 1971 and 1980, Cuba experienced extensive outbreaks of swine flu that inflicted significant damage on the country's pig farming. In 1971, over 740,000 animals had to be slaughtered, and in 1980, nearly 300,000. The Cubans claimed that the USA was behind these outbreaks, but little evidence supports this. In 1971, the outbreak of swine flu was attributed to virus-contaminated meat from abroad. In 1980, it is believed that the virus arrived with animals from refugees from Haiti stranded in Cuba after failed attempts to reach Florida. The USA was likely not responsible for these epidemics. Recently, these allegations have been revived but without convincing documentation.[174]

Tobacco production is important in Cuba. In 1979 and 1980, tobacco plants were severely attacked by a particular fungus, causing 25% and 90% of production to be lost respectively. Again, the Cuban authorities claimed that the USA was behind this. However, it is well known that spores from this fungus can be carried by the wind over large distances. It is therefore likely that the spores reached Cuba in this manner from other places in the West Indies where the fungus is found.[173]

Sugar production is also a vital part of agriculture in Cuba. Sugar cane can be attacked by a fungus that causes disease – sugar cane rust. In 1979, such an outbreak occurred in Cuba. Once again, the Cubans pointed to the USA, but the background for the outbreak is believed to be fungal spores blown in from the Dominican Republic.[173]

Dengue fever is an infectious disease caused by a virus that is widespread throughout the world, particularly in tropical areas of Africa and Southeast Asia. The infection is usually mild but can occasionally be severe with significant mortality. The virus is transmitted by a specific species of mosquito.

In 1980–81, Cuba was hit by a massive dengue fever epidemic, with nearly 350,000 Cubans infected and 158 deaths. As usual, the USA was accused of being behind the epidemic, without any evidence being presented. Cuba had experienced dengue fever both before and after 1980, without the USA being implicated. The most likely background for the 1980 epidemic is that Cuban soldiers who had fought in Angola brought the virus home, as also occurred with the HIV virus. The soldiers likely also brought back a virus causing an acute, painful eye infection. Again, the Americans were blamed.[173]

It is likely that all the mentioned cases of alleged bio-attacks from the USA are propaganda aimed particularly at the domestic market, where it was a matter of maintaining revolutionary fervour. The accusations against the USA were not investigated thoroughly by the Cubans.

However, in one case, the Cuban authorities went considerably harder. Now, it concerned an insect – *Thrips palmi.*[173]

From 1989, this pest had established itself on a long series of islands and countries in the West Indies and in December 1996, it was detected in Cuba for the first time. The Cubans alleged that this insect was released from an American plane that lawfully flew over Cuba in October. From another nearby plane, it was observed that the American aircraft released an unknown substance 'in the form of a whitish or greyish mist'. The Americans explained this as the camouflaged American plane having released a smoke cloud as a warning to the Cuban aircraft to prevent a collision.

Cuban authorities reported this case to the Biological Weapons Convention (BWC), which handled the accusations against the USA in its own meetings allowed by the convention. Nonetheless, no expert investigation was initiated, and there was also no consensus backing the conclusion from the BWC. The majority of states dismissed Cuba's accusations as unfounded, a few countries stated it was impossible to determine the issue, while North Korea alone concluded that the USA was guilty.

This case illustrates, as mentioned previously, the BWC's inability to intervene actively in questions of bioweapons use.

Operation Denver – KGB Campaign on the Origin of the HIV Virus

In December 1980, a young man was admitted to a hospital in Los Angeles with a remarkable clinical picture. He was affected severely by a fungal infection that is usually only seen in patients with severe immune deficiency. Over time, he developed a series of new infections of the same type. In the following months, more and more similar cases appeared among men on both the east and west coasts of the USA. They all shared one commonality: they were homosexual or bisexual men, and they had previously been completely healthy, without any diseases or treatments that could lead to immune deficiency.[110]

This frightening, newly emerging epidemic attracted significant attention. The media sensationally proclaimed that a 'gay plague' was now spreading in the USA. However, this designation proved to be inaccurate. It became clear in a short time that the epidemic was not only affecting homosexual and bisexual men but also patients who received blood products, immigrants from Haiti, drug users, and female sexual partners of men with the new disease. Even small children were affected.

Eventually, this new epidemic spread from the USA and affected all inhabited continents –this was a true pandemic.

It was confirmed quickly that the patients had a severe immune deficiency with a specific pattern. It primarily affected and destroyed a crucial type of cell in the immune system – the CD4-positive T-lymphocytes – which coordinate many functions of the immune system. As more and more of these cells are destroyed, the immune system collapses, leaving the patient vulnerable to a wide range of infections and some cancers.[110]

The new epidemic received the descriptive name AIDS, which stands for acquired immunodeficiency syndrome.

Why had this particular immunodeficiency disease struck these patients? A wide range of theories about the cause of AIDS were proposed – from relatively rational, medically plausible hypotheses to the most absurd conspiracy theories.

Some believed the cause must have been a microbe from outer space, similar to the Andromeda virus in the disaster film *The Andromeda Strain* based on a thriller by the author Michael Crichton. Others thought that the particular sexual lifestyle in homosexual environments was the explanation, either as the Lord's punishment for sinful ways or because intense sexual activity simply exhausted the immune system.

However, in the USA, some homosexuals and African Americans suspected that this was a man-made virus developed specifically to target them.

In any case, most people – also among researchers – believed that the cause of AIDS must be a new virus. In 1983–84, the retrovirus we now call HIV was identified after an intense research effort, particularly in the USA and France. But conspiracy theories continued, now connected to this newly discovered virus.

KGB and Stasi Take Action

During this time, the Cold War was still in effect. In these years, the KGB was particularly active in propaganda and psychological warfare against the West – a campaign they described as 'active measures'. Early in the AIDS pandemic, the KGB saw an opportunity to use it in a large-scale propaganda campaign against the USA. The campaign the KGB launched was perhaps one of the most successful in the history of this intelligence agency.

As early as 1983, the KGB made the first test of its new campaign in a smaller Indian newspaper, *Patriot*, that was Soviet-controlled. The paper published a letter from an anonymous sender – allegedly a prominent American scientist – claiming that AIDS was caused by a virus created at

Fort Detrick in the USA as part of the Americans' bioweapons programme. The letter attracted little attention but in 1985, the KGB struck in earnest. Now, all Soviet mass media published the claims that the HIV virus was created by the Americans. The KGB also engaged the secret services in the Soviet satellite states in this propaganda offensive.[175,176]

Particularly, the East German intelligence service, the Stasi, was an active partner in the campaign. The Stasi code-named the operation Denver, possibly due to a confusion between Detrick and Denver. The documentation for this campaign was found by a German historian in the archives of the Bulgarian intelligence service, which had been asked by the Stasi to participate. Unfortunately, the Stasi's own archives regarding this have been destroyed. It was the Stasi's sub-department for intelligence directed abroad – HAVA (*Hauptverwaltung Aufklärung*) – that was responsible for this propaganda operation.[175,176]

It is interesting that several prominent Soviet scientists publicly rejected the claims that HIV was man-made, despite the KGB's vigorous efforts. One of them – virus researcher Viktor Zhdanov – once responded to questions about this: 'It's a ridiculous question. Perhaps it was Martians.'[177]

Only one scientist could be used by the Stasi in this campaign, the Soviet-German retired biologist Jakob Segal, who was then living in East Germany. Without having special expertise in the area, Segal became fully convinced that the virus stemmed from Fort Detrick. Together with a colleague and his wife Lilly, who was also a researcher, Segal presented his views on the virus's origin in a 47-page pamphlet titled 'AIDS – Its Nature and Origin'. In this pamphlet, Segal and co-authors claimed that the virus originated as part of the American bioweapons programme by military researchers combining two retroviruses: Visna virus, which infects sheep, and HTLV-1 virus, which can infect humans.[177]

Segal never published his views in any scientific journal. However, the aforementioned pamphlet became central in communist propaganda. Jakob Segal was portrayed as a 'French' scientist in order to create distance from the Soviet bloc. Stasi and KGB made significant efforts to arrange meetings between Segal and journalists from many countries, and he impressed many as highly convincing.

The propaganda offensive was also directed at the Third World. In September 1986, the Eighth Conference of Non-Aligned Nations was held in Harare, Zimbabwe. Representatives from over 100 countries attended the meeting, and there were also 20 participants from the KGB and four from the Stasi. The Stasi participants had laboriously photocopied a brochure based on Segal's pamphlet, which was distributed to the participants at the congress. It was titled: 'AIDS: USA-Home-made Evil, NOT from Africa'.[175, 176]

This message resonated greatly with African leaders, journalists, and intellectuals. Western AIDS research had now claimed that the HIV virus originated from Africa and resulted from a related virus – SIV – in monkeys that had, on several occasions, infected humans and subsequently adapted to humans as HIV. These research findings had provoked a storm of anger among Africans and were sharply rejected as expressions of racism. Attitudes were likely influenced by African Americans' suspicion that the virus was aimed at them. Conspiracy theories were further elaborated. In 1991, for example, it was alleged that the CIA had sent HIV-contaminated condoms to Africa.

After the conference in Harare, Segal's claims spread in the press across 25 African countries. They were also included in the final, official report from the conference. In the Stasi's offices, they likely chuckled with satisfaction.[177]

A new triumph was achieved by the East Germans when they managed to draw the bestselling Austrian author Johannes Mario Simmel into their campaign. Simmel's books were translated into 25 languages, and several were adapted into films. He had received awards and honours both in Austria and West Germany. During routine monitoring, the Stasi learned that Simmel was now planning to write a novel about the dangers of genetic manipulation and biological warfare. The Stasi then anonymously sent material about Jakob Segal's theories to Simmel, who eagerly swallowed the bait. He now wrote a 500-page novel about the biological arms race between the USA and the Soviet Union, in which Segal's claims were integrated uncritically as facts. The novel, titled *Doch mit den Clowns kamen die Tränen* in German, was translated into several languages, and a TV programme was also made based on the book.[177]

It is said that Markus Wolf, the Stasi's deputy and head of the HAVA department responsible for propaganda, had a stack of ten copies of Simmel's book on his desk.

The KGB further escalated the propaganda – not only against the Third World but also against Western countries. By influencing journalists, they managed to place allegations regarding the origin of the HIV virus in the media in many countries. The accusations against the USA were expanded to include allegations that the HIV virus was developed as a weapon against black people. The KGB reminded that the Americans had, in the 1700s, attacked the Indian population with the smallpox virus. They also claimed that the USA had helped Wouter Basson in South Africa develop bioweapons against black Africans.[177]

The propaganda offensive continued even after representatives of the Soviet authorities and the Soviet Academy of Sciences distanced themselves from the accusations of the USA's responsibility for the HIV virus in 1988.

In 1989, the KGB scaled back its HIV propaganda and focused on other themes within the area of 'active measures' against the West. However, the Stasi continued unwaveringly to disseminate Segal's claims, which he continued to expand, including accusations that the National Cancer Institute in Bethesda, Maryland, was heavily involved in HIV as a bioweapon. The same applied to Robert Gallo, one of the world's leading AIDS researchers, whom Segal now characterised as 'a huge gangster'.[177]

Segal's claims were featured prominently in newspapers and TV programmes in the United Kingdom, the USA, West Germany, and Sweden. Newspapers featured sensational headlines like 'Yankee business, not monkey business'. Segal himself led a crusade with lecture tours and interviews. The extent to which Segal was motivated by Stasi is unclear. However, it is highly likely that he had close contact with the Stasi representatives and originally received material from them. Eventually, he likely became wholly convinced that he was right, and he never changed his views. He died in 1995.[177]

The Conspiracy Theory Lives On

Neither Jakob Segal's death nor the dissolution of the Soviet Union and Stasi led to the cessation of the conspiracy theories regarding the origin of the HIV virus. It had no effect that Yevgeny Primakov, who had been deputy head of the KGB and had led the intelligence agency's foreign department, officially declared in 1992 that these claims were false and fabricated by the KGB.

Particularly, the allegations that the HIV virus is a bioweapon developed against black people have persisted, primarily in Africa, but also in Afro-American communities in the USA.

In Africa, the conspiracy theory gained strength as AIDS ravaged the continent south of the Sahara with greater and greater intensity.[177] For example, the president of Zimbabwe, Robert Mugabe, stated that AIDS is 'a white man's plot'. In 2004, the Kenyan biologist Wangari Maathai received the Nobel Peace Prize. The Nobel Committee apparently did not realise that Maathai had repeatedly praised this conspiracy theory, including statements claiming that the HIV virus was produced by 'evil-minded scientists' to 'wipe out the black race'. When the author of this book drew attention to this in newspaper articles, it caused consternation among Maathai's Norwegian supporters, who tried to downplay her unfortunate statements. The Nobel Committee assured that they would look into the matter, but no conclusion has been reached thus far.

The African belief in the HIV virus as a responsibility of the West was not diminished by the statements of the English journalist Edward Hooper, who claimed that the virus came to Africa with a polio vaccine where monkey cells had been used in production. This theory was refuted thoroughly.[178]

The fact that the conspiracy theories related to bioweapons originally found a foothold in Afro-American communities is likely due primarily to a profound and understandable mistrust of both the CIA and American health authorities. Under the auspices of the CIA, highly dubious experiments were carried out where biological and chemical weapons were tested on young, Afro-American drug users admitted to treatment clinics.

An even more significant impact was likely made by one of the major research scandals of recent times – the notorious Tuskegee Study. In 1932 in the small town of Tuskegee, Alabama, the American health service began a study of 399 poor African Americans diagnosed with syphilis. They were not informed of their diagnosis, only that they had 'bad blood'. The health service followed the patients for 40 years without administering any treatment for syphilis, even after penicillin became available in 1947. The scandal was exposed in the press in 1972. Many patients had then died from syphilis, several spouses were infected, and a number of their children were born with this severe infection.[13]

It is understandable that these experiences created a breeding ground for conspiracy theories in Afro-American communities, where they have thus persisted. As recently as 2005, a survey showed that 50% of African Americans believed that the HIV virus was man-made, and 25% believed it was created in government laboratories. Also, 12% believed that the virus was spread by the CIA – and 12% that it was an attempt at genocide against black people.[177]

All in all, it must be acknowledged that the KGB's propaganda offensive against the USA regarding the origin of the HIV virus was highly effective. It undoubtedly damaged the USA's reputation in large parts of the world. It has also increased distrust of Western science, which we continue to see examples of – even in Western countries.

This story also illustrates how difficult it can be to counter an effectively launched propaganda offensive such as Operation Denver.

Plague and Propaganda

It is not necessary to go back to the time of the Black Death to find examples of accusations that plague was used as a bioweapon. The plague bacterium – *Yersinia pestis* – is still on the list of potential bioweapons.

Plague has not yet been eradicated in the world and in several countries new outbreaks continue to be reported, often of modest scope. This also applies to India. In 1994, there was a plague outbreak in the state of Gujarat in West India, which was more extensive than in many years.

In early September, hospitals in the city of Surat – one of India's wealthiest cities with a population of 1.5 million in 1994 – saw an increasing number of admissions of patients suffering from severe pneumonia. It turned out that the patients had plague – specifically the particular form known as pneumonic plague. The hospitals became overwhelmed quickly, and only the very sick could be admitted. Plague had not been seen in Surat for 30 years.[179]

As we have seen, *Yersinia pestis* can cause three different disease forms: bubonic plague, pneumonic plague, and septicaemic plague.

All these forms of plague carry a high lethality if antibiotic treatment is not initiated quickly. What was striking in Surat was that most of the hospitalised patients had pneumonic plague, which is not the most common form of plague in natural epidemics. In retrospect, it is estimated that at least 1,200 patients were affected by plague. There were 54 fatalities, all with pneumonic plague.[179, 180]

Rumours arose quickly that Muslim terrorists had caused the epidemic by poisoning wells in Surat – a line of thinking reminiscent of the accusations against Jews during the Black Death.[179]

The background for the terror accusations was the tensions between Hindus and Muslims that have persisted since India's partition and have occasionally led to unrest. Right in the state of Gujarat, tensions were particularly high.

The terror rumours triggered panic in Surat and led to mass migration from the city. More than 500,000 left in a hurry, including several doctors. With this mass exodus, the plague epidemic also spread to other cities in India, including Delhi, Bombay (Mumbai), and Calcutta (Kolkata), however Surat was the hardest hit.

The plague epidemic in Surat also had extensive international repercussions that affected India in various ways. Tourism to the country fell by 25% in the autumn of 1994. A large number of international flights to India were cancelled, and at several international airports, flights from India were thoroughly disinfected. International investments in India fell significantly, and some countries banned all imports of Indian goods.[179, 180]

The healthcare system in Surat managed to contain the plague epidemic within a couple of weeks using aggressive, early antibiotic treatment. Rumours of Muslim terrorism eventually subsided. However, doubts began to arise as to whether there had really been a plague epidemic in Surat. Could classic plague have been confused with many of the other serious infectious diseases occurring in the area? During the fight against the epidemic, fairly simple diagnostic tests were employed with little use of the most reliable methods.[179]

Nonetheless, further investigations of patient samples using modern molecular biological tests later showed that the cases did indeed involve *Yersinia pestis* infections. Furthermore, studies of the bacteria's DNA indicated that the epidemic in Surat was caused by a new variant of the plague bacterium that had slightly different properties than previously known variants. The theories of bioterrorism flared up again. Was this new bacterial variant man-made as a biological weapon? The suspicion was directed against Pakistan and the USA.[179]

No documentation for the conspiracy theories has ever been presented. The Surat epidemic can be explained quite simply based on what we now know about plague and the characteristics of the plague bacterium.

Before the outbreak in Surat, there had been a plague outbreak in the small town of Beed, 400 kilometres from Surat. At the end of August, 460 cases of an infection indicating classical bubonic plague were gradually discovered. The diagnosis was confirmed in submitted samples, albeit as in Surat with simple tests. Antibiotic treatment was initiated quickly, and no deaths were recorded.[179, 180]

In several places in Asia – including India – *Yersinia pestis* is permanently present in so-called plague reservoirs in wild rodent colonies. Changes in rodent populations due to climate change or natural disasters like earthquakes can bring infected rodents into close contact with humans. The plague bacterium can then be transmitted to humans via fleas, resulting in bubonic plague.[13] This had occurred in Beed, likely as a consequence of an earthquake a while beforehand.[179]

Many workers in Surat originated from Beed and maintained constant contact with their home town through frequent visits. In Surat, they lived clustered together and were relatively isolated from other ethnic groups in the city. It was precisely in the slum area with residents from Beed that the first plague cases appeared.[179, 180]

The plague epidemic in Surat in 1994 can, therefore, be explained based on the plague bacterium's characteristics and behaviour in nature, along with specific social conditions in the city. That there were no fatalities among the bubonic plague patients in Beed may be due to the particular bacterial variant being somewhat less virulent – less 'angry' – than commonly occurring variants of *Yersinia pestis*. If the bacterium had been modified in the laboratory as a bioweapon, one would have expected it to be produced as more virulent.

Bioweapons and the War in Ukraine

On 24 February 2022, Russian forces invaded Ukraine. Since the invasion, the war has been ongoing with full force and has resulted in a significant number of deaths on both sides. The invasion marked an escalation of the conflict between Ukraine and Russia that has persisted since 2014 after Russia's occupation of Crimea. There are currently no signs that the war will end in the near future.

That the truth is the first casualty in a war is a somewhat worn-out quote but, like all clichés, it has a kernel of truth. The war in Ukraine illustrates this very clearly. Alongside the conventional war, there has also been another, equally vigorous war, the extent of which we have scarcely seen in any earlier military conflict – a propaganda war between Russia on one side and Ukraine and the USA on the other.

From the Russian side, there have been repeated accusations that several state laboratories in Ukraine, with American support, particularly from the Pentagon, are working on the development of bioweapons – highly pathogenic bacteria and viruses. The Russians allege that Ukrainian authorities ordered the destruction of highly pathogenic microbes in these laboratories shortly after the invasion.[181, 182]

The Russian allegations have been categorically dismissed by both the White House and the Pentagon, which in turn has suggested that this could be a so-called false flag operation signalling the potential use of biological weapons by Russia.[183]

Americans acknowledge that they are involved in several state biolaboratories in Ukraine, but maintain that the USA's engagement is part of the Cooperative Threat Reduction Programme (CTRP), which is administered by the Pentagon. As described previously, this programme was established in 1991 to ensure that nuclear, biological, and chemical weapons in the former Soviet Union did not go astray but were responsibly stored or destroyed. The work in Biopreparat's many laboratories across the old Soviet empire thus fell under the CTRP. This included several laboratories in Ukraine, and collaboration between the CTRP and Ukraine was further strengthened by an agreement in 2005. The Pentagon also sponsored many Western researchers in these laboratories. The USA asserts that the research in these laboratories focuses on the prevention and monitoring of natural and man-made epidemics in both humans and animals, and that it exclusively concerns bio-defence and not the development of bioweapons.[184]

Robert Pope, who now leads CTRP, categorically denies that these laboratories in Ukraine are conducting research on bioweapons, although some of them were engaged in this during the Soviet period. He claims they conduct legitimate, 'peaceful' research aimed at detecting new microbe threats, such as swine flu. According to Pope, several of these laboratories undoubtedly still have dangerous microbes stored in freezers, some of which likely originated from Biopreparat's work with bioweapons. The CTRP reportedly requested previously that Ukrainian authorities destroy these microbes, but this work had not begun when the war in Ukraine broke out. WHO also requested in March 2022 that Ukraine destroy especially dangerous microbes to prevent their spread 'either by accident or deliberately'.[185]

Pope's statements from February 2022 may well align with reality but not everyone is equally convinced, neither within Russia nor outside. It is a fact that the USA, over many years, actively developed bioweapons, and in 1969, when Nixon officially announced that this work should cease, the country had one of the most extensive bioweapons programmes the world has ever seen. It cannot be ruled out that the USA still has such a programme, part of which may also be conducted in other countries. It has attracted some attention that Robert F. Kennedy Jr, who was originally a candidate in the presidential election of 2024, supports these claims.[187]

Scepticism towards the official American version regarding the USA's involvement in Ukrainian laboratories was fuelled by statements made by Victoria Nuland, the US Deputy Secretary of State, during a Senate hearing on 8 March 2022: 'Ukraine has biological research facilities, and we are now very concerned that Russian forces may try to seize them.' She further stated that 'we are working with the Ukrainians to prevent this research material from falling into the hands of the Russians'.[188] What kind of material could this involve? Russian microbiological laboratories are advanced and, of course, have access to all kinds of normally occurring microbes, including highly pathogenic ones. Were Nuland's statements, in reality, an involuntary admission that compromising material related to bioweapons existed in these laboratories?[189]

Western media appears commonly convinced that all Russian allegations regarding the war in Ukraine are propaganda lies and disinformation. This also applies to the claims regarding bioweapons. However, it is a somewhat under-reported fact that there is also propaganda coming from the Western side. We currently do not have sufficient information to take a definitive stance in the propaganda war over the biolaboratories in Ukraine. Regardless, the situation gives cause for concern. If laboratories with potentially dangerous microbes in freezers are damaged by military actions, power outages could lead to the microbes thawing and leaking, resulting in dangerous epidemics. The microbes could also, in a chaotic wartime situation, fall into the hands of extreme Ukrainian groups that could attempt to weaponise them against Russian invasion forces.[189]

Russia has consistently maintained its accusations against the USA and Ukraine, and the White House has also been drawn into this conflict. There are certain grounds for the belief that former President Joe Biden's controversial son, Hunter Biden, has significant financial interests in a private American company involved in research at Ukrainian biolaboratories. Additionally, he is said to have engaged in active lobbying for this company in Ukraine.[190]

In the first half of 2022, Russia advanced its accusations in the UN General Assembly and repeatedly in the Security Council. By referencing Article V of the BWC, Russia was able to hold a separate meeting in the convention in September 2022 concerning bioweapons in Ukraine. The last time such a meeting was held was when Cuba accused the USA to the BWC in 1997 over allegations concerning the spread of the pest insect *Thrips palmi*. As we recall, this did not lead to any concrete results, nor did Russia's bioweapon accusations against the USA.[191]

Russia went further and referred to Article VI of the BWC, which connects the UN Security Council and allows for the establishment of a specific investigative commission. This provision has never been used before. The Security Council meeting in October 2022 concluded with only China supporting Russia among the permanent members, and only a few of the non-permanent members. In other contexts, China has also supported Russia's claims regarding US activities in Ukraine.[191]

As long as neither Russia nor the USA can present convincing evidence for their claims, this propaganda war will continue. However, it has had one positive result. The threat that biological weapons represent – a threat that is significantly under-reported in public discourse – has once again come to the forefront. The fruitless meetings in the UN and BWC concerning this dispute have also demonstrated that the BWC must now be supplemented with clear provisions ensuring compliance with the treaty, along with effective oversight of suspicious activities. Unfortunately, no immediate solution to this is in sight.

Monkeypox Epidemic – Natural Outbreak or 'Monkey Business'?

As the world was just beginning to recover from the worst effects of the Covid-19 pandemic, a new virus emerged on the scene. From early May 2022, there was a growing number of cases of monkeypox virus infection – especially in Europe and the USA. The cases appeared almost simultaneously in a number of countries and eventually, all continents were affected by the epidemic.[192] By August 2023, nearly 90,000 cases had been reported from 113 countries, with 153 deaths. The virus was given a new name by WHO – *mpox virus* – ostensibly to prevent misunderstandings about its origin.

Many have wondered about the background of this epidemic involving a virus that previously had only caused infections in certain African countries. Why did this epidemic surface precisely now? Shortly after the outbreak, a number of conspiracy theories emerged – ranging from somewhat unlikely to more plausible explanations.

There is still much we do not know about the monkeypox virus. However, current knowledge allows for a sober assessment of the claims regarding the virus's origin.

The common denominator for most conspiratorial claims is that individuals with malicious intent have unleashed the virus on the global population. In these theories, the USA emerges as the primary actor. Here, too, Russia has been particularly active in accusing the USA. Igor Kirillov, representing the Russian Ministry of Defence, declared to Russian media that the virus originated from American military biolabs in Nigeria as part of bioweapon research.[193]

Proponents of this view have one main basis for their belief: the American organisation Nuclear Threat Initiative (NTI) organised a conference in Munich in March 2021, during which a panel of 19 experts was asked to assess global pandemic preparedness in the event of a hypothetical outbreak.[194] They were presented with a fictional terrorist attack involving a monkeypox virus that had been modified in the laboratory to make it more dangerous – more virulent. According to this hypothetical scenario, the pandemic was set to break out in May 2022. The context of NTI's theoretical pandemic exercise is quite plausible. Some epidemic experts have indeed considered monkeypox virus a potential new epidemic threat for several years, yet researchers have not shown much interest in this until now. But they should have.

Monkeypox is a classic example of the emerging infectious diseases that have surfaced over the last 50 years.[13] In most cases, this does not involve a truly new microbe, but rather changes in the environment and ecology that have brought an already existing microbe into contact with humans. These changes are very often human-induced. Deforestation, war, and unrest have likely increased contact between humans and the animal species that carry the monkeypox virus – primarily rodents. Increased consumption of so-called bushmeat – meat from wild animals such as rodents and primates – has been significant, as such meat is part of the diet in several African countries.

The monkeypox virus belongs to a large virus family – *Orthopoxvirus* – where we also find the related, but much more dangerous, smallpox virus.[195]

A significant factor behind the increase in monkeypox infections seen in Africa over the past decades has been the cessation of smallpox vaccinations since around 1980, when the virus was considered eradicated globally. Vaccine-induced immunity against smallpox also provides considerable immunity against monkeypox. Most cases of monkeypox in the first years after 1980 were among unvaccinated children. As fewer and fewer people in the affected countries have immunity from smallpox vaccination, monkeypox has increased in frequency alongside the rising average age of patients. This infection has become a significant health problem in Central and West Africa.[195]

Until 2022, only a modest number of monkeypox cases had been reported in Europe and the USA, where the virus was introduced from Africa through travellers or via imported animals. However, a larger outbreak of this virus outside Africa has always posed a risk. Therefore, NTI's choice of the monkeypox virus for their pandemic exercise was, in itself, reasonable.

Those who doubt that this epidemic has a natural background emphasize that the NTI's hypothetical pandemic exercise presumes that the virus outbreak would occur in May 2022 – a timeline that coincides with the actual outbreak. It must be conceded that this is indeed a striking coincidence. That we are faced with a demonic plan to trigger a pandemic, however, does not align well with the fact that NTI publicly released its entire scenario, including the proposed start date in May 2022, back in November 2021.

Another point that conspiracy theorists highlight as suspicious is that the USA possesses a significant stockpile of smallpox vaccines despite the fact that smallpox was declared eradicated in 1980. Have the Americans perhaps prepared for a monkeypox pandemic? The more prosaic explanation for this, however, is that the smallpox virus is on the list of possible microbes used in bioterrorism, and the vaccine stockpiling is part of preparedness against biological weapons.[196]

Some conspiracy theories assert that the smallpox virus causing the current epidemic has recently been modified in a laboratory to change its mode of transmission and make it more contagious. They emphasize the notable characteristic of the non-African epidemic: that 98% of those affected in Europe and the USA are homosexual and bisexual men. There was suspicion of sexual transmission in 95% of the cases. The virus being detected in the semen of most who have been examined also suggests that sexual transmission is likely, although close bodily contact is also important.[192]

However, decades of experience from the African epidemic show that the virus appears to have changed naturally in recent years regarding modes of transmission. In the early years, human-to-human transmission was relatively uncommon – it occurred in Central Africa, but hardly ever in West Africa. In Nigeria, epidemiologists believe that a change occurred in 2017 when a large outbreak of monkeypox broke out there. While most cases occurred previously in rural areas, the virus now concentrated in the urban environment. There were also indications that the virus had begun spreading among humans – also through sexual contact.

This may provide the key to understanding the outbreak outside of Africa. The virus in the ongoing epidemic is the variant that occurs in West Africa.

Several research groups have shown recently that this virus has undergone several natural genetic changes over the past five years. Thus, there is no reason to suspect that the virus is laboratory manipulated.[197]

The monkeypox outbreak in Europe and the USA bears several similarities to the first phase of the HIV pandemic. The relatively non-contagious HIV virus, with origins in Africa, was introduced into Western countries within certain communities of homosexual and bisexual men where specific sexual behaviour patterns led to rapid spread. As is known, the HIV virus eventually spread to some other groups as well. The same is now occurring with monkeypox. Cases in children and women have already been reported.[196]

Some conspiracy theorists have emphasized that the epidemic broke out almost simultaneously in several countries and suggest that this may indicate a terrorist attack. However, this pattern likely results from increased international travel associated with participation in large events featuring 'sex-on-site' activities, saunas, and 'sex parties'.[192]

It seems unlikely that American authorities would specifically want to target homosexual men. Finally, it seems quite improbable that one would choose the relatively low-virulent monkeypox virus, with a lethality rate of 3–6%, as a biological weapon.

Overall, all aspects of the monkeypox virus epidemic that broke out in 2022 can likely be explained without resorting to conspiracy theories about laboratory-produced biological weapons and terrorist attacks. A completely different matter is that this virus, like other viruses in the Orthopox family, may in the future be modified to become more dangerous and deadly as a biological weapon.

Allegations about Taiwan and Bioweapons

For decades, there has been significant tension between the People's Republic of China and Taiwan. Taiwan's claim for independence is flatly rejected by China, which has never deviated from its demand that the island republic must be integrated into the 'motherland'. Taiwan has full support from the USA, making this conflict a potential threat to world peace.

It drew considerable attention when one of Taiwan's largest newspapers reported in July 2022 that it had secure information about a meeting in the

South China Sea between Taiwan's ruling party and representatives from the USA. Reportedly, the Americans requested Taiwan to build a high-security laboratory ('biosafety level 4') to develop biological weapons. The newspaper followed up a few days later with alleged photographs of documents from this meeting. The claims have since been actively spread, particularly in Chinese-controlled media.[198]

Taiwan has developed significant expertise in biotechnology, and over the years the country has been listed among states that may have an active bioweapons programme.

Regarding the recent claims, both Taiwan and the USA have firmly denied that the USA has initiated any bioweapons development in Taiwan. Representatives of the Taiwanese government assert that they wish solely to develop defences against biological weapons and not to produce them.[199]

There has also been significant doubt raised about the authenticity of these revealing documents. It has been pointed out that neither the wording nor the presentation bear the characteristics of official government documents. It is therefore not unlikely that this is a propaganda offensive from China against Taiwan in the ongoing conflict between them.[198]

From Taiwan's perspective, there is likely fear that China could use biological weapons in the event of an invasion. Therefore, it is understandable that the island republic is aiming to strengthen its bio-defence. This is most likely the background for the sensational headlines about the meeting between Taiwan and the USA.

This episode is another example of how allegations of bioweapons development and possible use of such weapons are popular in propaganda contexts.

Was the Covid-19 Virus Developed as a Bioweapon?

The two main hypotheses regarding the origin of the Covid-19 virus are, as mentioned previously, that it involves a zoonosis or that the virus has 'leaked' accidentally from the Wuhan Institute of Virology as part of gain-of-function research on coronaviruses. As we have seen, there is no definitive documentation for either hypothesis.

Some have also proposed a third hypothesis, namely that the virus was developed as a bioweapon. We know that Chinese military medicine conducts extensive microbiological research at the WIV, and there is much to suggest that China has a bioweapons programme.

In the USA, a Republican senator, Tom Cotton, has put forth this bioweapon theory, but no intelligence agencies or researchers have regarded it as likely so far.[200] It has attracted attention that Robert F. Kennedy Jr has supported the theory.[201]

Recently, a Chinese researcher who has worked at the WIV stated in an interview that Covid-19 had indeed been developed by China as a bioweapon and tested various variants on unsuspecting individuals in society – including Uighurs in Xinjiang province and participants at a congress in Wuhan in 2019. However, this involves only claims from a single researcher who has not presented any evidence for this.[202]

This is a particularly delicate aspect of the bioweapon hypothesis: Anthony Fauci, who has significantly supported the USA's funding of coronavirus projects at the WIV, has played a central role in the USA's bio-defence programme since 2004. Given the tense relations between the superpowers USA and China, it would be surprising and highly unlikely that Fauci would have approved and supported collaboration on bioweapons development. Nevertheless, it is striking – and also highly questionable – that NIAID under Fauci's leadership has supported risky gain-of-function experiments at the WIV where Chinese military doctors also work. This is likely an important reason why he has steadfastly denied that he and NIAID funded GOFR at the WIV.

Furthermore, it would also be surprising if Chinese scientists intentionally released the virus within China's borders, given the enormous consequences the pandemic had for Chinese society as well as for the rest of the world. Unless new and more solid data emerges, the hypothesis of Covid-19 as a bioweapon must thus be considered unlikely.

It cannot, however, be completely ruled out that work with various coronaviruses at the WIV may have been part of a Chinese bioweapons programme. As mentioned, Chinese military medicine is heavily involved in this laboratory. Recent reports have indicated that researchers from the

Chinese military have shown that a laboratory-modified coronavirus related to the Covid-19 virus is 100% lethal in mice and may possibly infect humans.[203]

The background of the Covid-19 virus remains unresolved. Unless China wishes to collaborate on this issue, we may never obtain a definitive answer regarding the virus's origin.[204]

Chapter 15

A Deadly Threat? – How Real is the Danger from Biological Weapons?

Will states use biological weapons?

Biological weapons have a long history. Today, it is believed that several countries possess such weapons. However, since biological weapons have been condemned by the international community in recent years, the information is uncertain, and effective control mechanisms are lacking.

Along with nuclear and chemical weapons, biological weapons are considered 'weapons of mass destruction'. Despite this, the silence about such weapons is deafening in the media, among politicians, and in public debate in general – unlike the strong concerns regarding the possible use of nuclear weapons in, for example, the Ukraine war. In the Norwegian government's report on health preparedness dated 24 November 2023, bioterrorism is mentioned only once in parentheses in a side note. The possibility of military use of biological weapons is not discussed at all.[205] Does the apparent lack of interest in biological weapons mean that they are now dismissed as a serious threat and are considered almost a historical phenomenon without relevance to our time?

Although the USA, the Soviet Union, and several other countries had extensive biological weapons programmes, such weapons have only been used definitively in warfare by two countries, namely Germany during World War I and Japan before and during World War II. This does not mean that biological weapons are no longer relevant in a military context; there are several possible reasons why these weapons have so far not been used beyond the mentioned examples.

As we have seen, there are several international agreements aimed at preventing the use of biological weapons, primarily the Biological Weapons Convention (BWC). These are mainly based on the notion that such weapons are considered particularly abhorrent compared to conventional weapons. Not everyone has viewed this as entirely rational; many have argued that it makes little difference whether soldiers or civilians are harmed or die from conventional or biological weapons.

However, it is likely that the moral attitudes underpinning international conventions like the BWC play a certain role. Knowledge that a state had employed biological weapons in a military conflict would probably be met with strong condemnation and massive sanctions against the 'sinner' from other states.

In my opinion, however, it is naive to believe that states base their military build-up and possible warfare on anything other than strict practicality. If biological weapons in some form are considered effective and useful as a supplement to conventional weapons, they will likely be pursued. But due to the strong condemnation of such weapons, this pursuit would occur in deep secrecy and be officially denied. This has been seen before, and it is likely the case today – regardless of loud, official assurances that investment is only in non-offensive bio-defence.

There is thus little doubt that biological weapons are still relevant. If they are seen as useful in a military conflict, they will likely be used. As Cicero asserted: *Inter arma silent leges* – 'In war, the laws are silent'.

As an argument against the biological weapons threat, it is often claimed that biological weapons used in war are difficult to manage, have unpredictable effects, and pose a significant risk even to the side that employs them. Japanese experiences in Manchuria and China show both that the military utility of their biological weapons was probably limited and that a significant number of Japanese soldiers were affected by them. However, the Japanese biological weapons were rather primitive, and their practical use – especially the methods of dissemination – was not sophisticated compared to the results achieved by several countries later in the twentieth century.

Thus far, vaccines have only been developed against a few biological weapons. With the significant advances in vaccine technology in recent years, it will become possible to produce vaccines against many more biological weapons. A military force wishing to employ biological weapons would therefore be able to protect its own troops through vaccination.

Fifty years ago, when the USA's bioweapons programme was at its most extensive before it was officially terminated, numerous field trials with such weapons were conducted, showing that they could be used with high precision and likely massive effect over larger distances. Similar experiences were seen in some other countries, such as Great Britain and probably also the Soviet Union.

The arguments for downplaying the continued relevance of biological weapons are thus untenable.

Bioterrorism – An Increasing Threat

What about terrorists? We have discussed previously possible reasons why we have not seen more terrorist attacks with biological weapons than quite primitive attempts, such as those made by the Rajneeshi and Aum Shinrikyo cults.

However, much has happened in recent years that has increased the risk of terrorists using biological weapons.[113, 207, 208] Over time, many more have acquired knowledge in microbiology and biotechnology that could make them capable of working with microbes as weapons. It is also not difficult to obtain microbes for this work from both commercial and non-commercial suppliers. Furthermore, there are detailed 'recipes' for necessary laboratory procedures available on the internet and in journal articles. Additionally, there are few requirements for laboratory space or advanced laboratory equipment for terrorists wanting to invest in biological weapons. This allows such illicit activities to easily go under the radar. Moreover, entirely new methods for disseminating biological weapons have become available with the explosive development of drone technology.

Extensive biological weapon attacks across larger land areas with high death tolls would likely still require expertise and resources that only state actors possess when it comes to the professional production of 'weaponised' microbes that can be stored over time and have significant stability. However, it is likely an exaggeration when a social actor, such as Bill Gates, claims repeatedly that a bioterror attack could claim millions of lives.[209] Nonetheless, there is little doubt that bioterrorists could cause significant damage with simpler weapons.

Another matter is that terrorist groups – backed by states with resources for producing sophisticated biological weapons – could gain access to such weapons that would significantly increase their damage potential.

In addition to the conditions already mentioned, there has been quite a dramatic development in research and technology that will clearly increase the threat from biological weapons in the years ahead. This applies to both state and terrorist use of such agents.

In summary, there is no reason to dismiss the threat from biological weapons, either on the battlefield or in terror attacks. In addition to the conditions already mentioned, there has been a dramatic development in research and technology that will clearly increase the threat from biological weapons in the years to come.

Chapter 16

The Price of Progress – Biological Weapons and Technological Advances

In recent years, there has been rapid development in several fields of research and technology. This development not only enables the production of entirely new biological weapons but also provides new opportunities for their effective dissemination. These advances significantly illustrate the 'dual use' problem, as they are undoubtedly of great benefit in many societal areas while simultaneously being able to be misused to produce new types of biological weapons with even greater potential for harm than conventional ones. Let us look at the most important of these advancements.

Biotechnology and Synthetic Biology

It has been said, not without reason, that the twentieth century was the century of physics, while the twenty-first will be the century of biology.[210] After Watson and Crick revealed the secrets of the DNA molecule and its significance for genetics in 1953, there has been explosive development both in fundamental research in this area and in the practical application of research results.

A range of methods have been developed that allow manipulation of DNA molecules to change their functions in cells and organisms. A landmark discovery was the method known as CRISPR, which allows targeted editing of the DNA molecule in a more efficient manner than previous methods allowed. We now operate with the term synthetic biology. This is not defined precisely,[129] as synthetic biology is today used as a collective term for the revolutionary use of biotechnological techniques to create entirely or partially artificial biological systems with new properties – ranging from simple molecules to entire organisms.[210, 211] Synthetic biology offers far greater possibilities than older gene manipulation techniques, which typically led to less-profound changes in function. Therefore, synthetic biology has been called 'gene editing on steroids' to describe how groundbreaking this field actually is.

With synthetic biology, it is now possible to make changes in the genetic material (DNA and RNA) of viruses and bacteria that can radically alter the properties of these microbes or even create entirely new organisms. This is highly relevant for the development of biological weapons. The numerous possibilities offered by synthetic biology could lead to the creation of new and far more dangerous biological weapons than those that have garnered attention previously and against which defences have been planned.[212]

More Dangerous Microbes

Genetic manipulation can make existing microbes more dangerous as biological weapons. This can happen in many different ways.[213]

One could increase a microbe's ability to infect new animal species.[213] For example, a virus that in nature can only infect certain species might be altered so that it can also infect humans. This was the danger associated with experiments using the avian influenza virus H5N1, which sparked the debate around gain-of-function research. Similar experiments have already been conducted with relatives of the Covid-19 virus, SARS-CoV-1, and MERS virus. These coronaviruses have a much higher fatality rate than the Covid-19 virus but are significantly less contagious. A substantial increase in contagiousness could make these viruses terrifying biological weapons. Genetic manipulation can also increase the reproductive capacity of viruses, which could lead to higher transmissibility.

In the ongoing debate about the origins of the Covid-19 virus, some have claimed that it not only originated from a laboratory in Wuhan but that it has been genetically manipulated to transmit more easily between humans.

A microbe's virulence – the ability to cause disease – can also be increased in the laboratory.[213] Virulence is often determined by multiple genes, and the result of laboratory manipulation of these can be difficult to predict. A very convincing example of how a virus can be made much deadlier is the experiments conducted in Australia in 2001 on mice with the mousepox virus – a relative of the smallpox virus.[214] The researchers, who had entirely different intentions for the experiment, made the following discovery: when they injected mousepox virus that they had 'boosted' with a new gene to increase antibody production against the virus, all the mice that normally did not get particularly ill from this infection died. It turned out that the new, deadly virus had disabled critical parts of the animals' immune systems. If the already highly lethal smallpox virus were to be manipulated in the same way, the result could be a horrific viral weapon.

There are also numerous possibilities for increasing the virulence of bacteria. Many bacteria produce toxins that are essential for their ability to cause severe illness, such as diphtheria, tetanus, and botulinum toxin. Through genetic manipulation, one can increase toxin production and also develop it in normally non-producing bacteria.[213]

For both viruses and bacteria, there are many ways to change the microbe's properties so that it can go under the radar of the infected individual's immune system. The harmful effects of the virus would then increase.

In a biological weapon attack, it is, of course, important to identify the microbe in question as quickly as possible to initiate treatment and other countermeasures. This may become more challenging if the microbe is manipulated so that it can no longer be detected by commonly used methods, which in both viruses and bacteria are often based on the so-called PCR technique. Such 'invisibility' of the virus is possible with today's methods.[213]

While we currently only have medications effective against a handful of viruses, we have antibiotics against a vast number of bacteria. A new approach in biological weapon development with terrifying implications would be to genetically manipulate microbes to make them resistant to the most common medications. This is particularly true for bacteria where genes that can cause antibiotic resistance are known and can be manipulated. The Soviet Union was already working on this in its bioweapons programme. Antibiotic-resistant bacteria with high mortality rates – such as the plague bacterium – would be exceedingly dangerous as weapons.

In attempts at extensive biological weapon attacks aiming to impact a large number of people over broad areas, the effectiveness will largely depend on the microbe having a relatively high degree of stability, meaning that it is not quickly destroyed by environmental factors like temperature, ultraviolet radiation, and humidity. It will likely become possible to stabilise both viruses and bacteria so that they are better suited as weapons.[213]

Recreation of Known Disease-Causing Microbes

In synthetic biology, it is possible to create a microbe from scratch. This means that the microbe is made from the individual building blocks (nucleotides) of the microbe's genetic material –the genome – provided that the structure of this is known. Today, the genomes are determined and publicly available for many microbes.

By 2020, the genomes of over 9,000 viruses had been published. The size of genomes varies greatly in the virus world; the larger the genome, the more difficult it is to synthesize. One of the first viruses built from scratch was the relatively small poliovirus in 2002.[215] Since then, larger and larger viruses have been synthesized. In 2005, American scientists managed to reconstruct the variant of the influenza virus that caused the Spanish flu 100 years ago.[216] In 2017, the horsepox virus – related to the smallpox virus – was constructed in this way.[217] Likewise, the smallpox virus could also be reconstructed by malicious actors.

There is reason to fear that both unavailable and hard-to-access viruses could now be synthesized for weapon use. This does not require more resources and expertise than a non-state bioterror group could possess.

Synthesizing bacteria from scratch where the gene structure is known is far more complicated than with viruses. The genomes that are known for many bacteria can likely be synthesized –but the challenge is to construct the entire bacterium, which is far more complicated than a virus.[213]

New Microbes Can Be Created in the Laboratory

An ambitious goal in synthetic biology is to produce entirely new microbes for useful purposes. By 'new', we mean microbes constructed from various gene segments from many different microbes so that the result is completely new – microbes that normally do not occur in nature. Here, there are great challenges and opportunities – but also significant dangers.

In 2010, American scientist Craig Venter succeeded in producing a new bacterium with a genome constructed in the laboratory using parts from other related bacteria. The resulting bacterium was named Synthia. It resembles existing bacteria in the *Mycoplasma* genus, some of which can cause infections in humans. This was a breakthrough in synthetic biology. However, Venter also recognised the potential for biological weapon development using synthetic harmful bacteria. Currently, the production of entirely new bacteria is so demanding that the threat is not considered significant at the moment, but this could change.[213]

Regarding the production of entirely new viruses, there has yet to be a breakthrough. The technical difficulties are considerable, but they will likely be overcome eventually. This could represent serious threats in the context of biological weapons. So far, such viruses have only appeared in science fiction literature, such as Richard Preston's previously mentioned sci-fi novel *The Cobra Event*. In this book, a terrorist releases a virus constructed from genes of the smallpox virus, a cold virus, and a virus that attacks insects.[71] The result is a horrifying microbe that spreads through the air and then attacks and destroys the brain within a few hours. One day, similar viruses may indeed be a reality. Scientists in the Soviet Biopreparat allegedly already had such chimeric viruses on the drawing board.

Gene-Directed Bioweapons Against Ethnic Groups

Biological weapons aimed at ethnic groups based on their specific genetic material belonged previously only in the realm of science fiction literature. In his book *The Day after Tomorrow* from 1941, Robert A. Heinlein introduced the idea of a race-specific biological weapon aimed at Asians.[219] The concept was revived by Jim Geraghty in 2020 in the book *Hunting Four Horsemen*, where a brilliant and unscrupulous scientist has developed a deadly virus that can target specific human genes.[220]

A main objection to the possibility of biological weapons targeting ethnic groups is that genetically distinct 'races' do not exist. Nevertheless, in geographically separated populations, it can be found that the frequency of certain genes differs markedly.[221] This includes genes and gene combinations that lead to reduced resistance to various microbes.[13] During the Covid-19 pandemic, for instance, it became clear that there were significant ethnic differences in the frequency of severe infection and death, which were at least partially genetically determined. Theoretically, it could therefore be possible to design biological weapons, such as a virus, that selectively target individuals with specific genes or gene combinations characteristic of an ethnic group.[222]

Such genetically selective biological weapons could be constructed to cause severe infection with the relevant microbe. However, they could also be designed to incapacitate central parts of the immune system, rendering affected individuals defenceless against infections in general. A naturally occurring example of microbes achieving this is the HIV virus, which targets a crucial cell type in the immune system – the CD4-positive T lymphocytes – and leads to immune deficiency.[110]

As we have seen, Wouter Basson in South Africa dreamed during his work on Project Coast of developing a 'black bomb' aimed at Black Africans. It is also interesting that Vladimir Putin, in 2007, imposed a ban on the export of genetic material from ethnic groups in Russia, possibly out of fear that it might be used to develop ethnic biological weapons. The same year, the influential Chinese general Zhang Shibo published a book on the scientific foundations of military strategy in which he mentions 'specific ethnic genetic attacks' as a possibility.[223] Recently, Chinese authorities have claimed that genetic weapons are currently under development in some countries, likely referring to the USA.[224] This could clearly be yet another example of propaganda but it illustrates that the fear of ethnic biological weapons is real.

With the explosive development in biotechnology and synthetic biology, the development of gene-directed ethnic weapons is likely a possibility in the years to come. This provides a new and alarming dimension to the biological weapons threat. It may also become possible to produce such weapons aimed exclusively at an individual's gene structure. If this occurs, it could be used as an assassination weapon against individuals. This technology was launched in the James Bond film *No Time to Die* from 2021, where the hero is infected using so-called nanobots – microscopic robots that in the movie are deadly for individuals with specific genes.[225] Such nanobots are currently under development for medical use.[226] They represent a new example of dual-use technology that could also be used as carriers of biological weapons in the future.

Reactivation of 'Sleeping' Viruses

One possible biological weapon strategy could be to mimic nature: one could exploit the fact that some naturally occurring viruses can cause a 'silent' infection in humans for many years before awakening and causing disease. This applies, for example, to herpes viruses that cause cold sores, the varicella virus that causes shingles, and the hepatitis C virus that causes liver disease. It may become possible to control the 'awakening' of such viruses externally with biological or chemical agents.[213]

In addition, it may also be possible to produce entirely new disease-causing viruses that can cause such 'sleeping' infections and later be activated.

Artificial Intelligence (AI)

The development of artificial intelligence (AI) has accelerated in recent years and has already had a significant, beneficial impact on many societal areas. At the same time, it has increasingly become clear that there are substantial risks of misuse and serious negative consequences from increasingly advanced AI systems. This includes, not least, the risk that AI will increase the threat from biological weapons in several ways.[227]

Instructions for Producing Known Bioweapons

First, AI-based language models – such as the recently developed ChatGPT – might might be utilized by bioterrorists. This could happen by disseminating comprehensive information about deadly microbes as possible biological weapons, as well as detailed instructions on how to produce them and how to obtain the necessary reagents. In a recent experiment with ChatGPT, it was demonstrated that it took only one hour for the system to provide information on four relevant pandemic microbes and how to acquire the necessary genetic material for their production unobtrusively.[228]

In working with complicated biological systems, 'tacit knowledge' may be necessary – insight that cannot be obtained from the internet but only through practical, hands-on work. Future more advanced AI systems may eventually be able to provide knowledge about techniques and procedures that are part of such tacit knowledge.[229]

Development of New Bioweapons

AI will also simplify the development of new biological weapons. In 2022, researchers conducted an experiment with an AI system instructed to identify potential toxic substances. In just six hours, the AI system had identified 40,000 different toxic chemicals – many of which were unknown previously.[230] Similarly, we must expect that AI systems could create new and more dangerous biological weapons with altered properties, as we have described earlier.

The many risks associated with the misuse of AI – especially in the context of biological weapons – make it imperative to establish effective regulations in this area. Intensive work is already under way on this but the challenges are significant.[231]

Drones and Death

In recent years, there have been significant advancements in drone technology. Drones are now employed in various societal areas where they can be used for surveillance, searching for individuals, and transporting goods. In agriculture, drones are already used to disperse pesticides in several countries in Asia. Many types of drones are commercially available on the open market, coming in many sizes. Over time, drones as small as insects have even been developed.

In the military domain, the introduction of drones has been nothing short of revolutionary and has become an essential part of modern warfare. This is clearly illustrated by the ongoing war in Ukraine, where drones have played an invaluable role both for reconnaissance and as weaponry: they can either release explosives from the air or crash into enemy targets with their explosives as so-called kamikaze drones.

Drones will, without a doubt, be almost ideal carriers for biological weapons.[232,233] Both in military contexts and for bioterrorists, drones represent a significant advancement as a delivery method for biological weapons. On the battlefield, drones capable of carrying substantial amounts of a suitable biological weapon could be used both to release such weapons from the air and as kamikaze weapons, where the release of microbes occurs on the ground when the drone crashes and is destroyed. Drones carrying explosives can be neutralised by being shot down. However, drones carrying biological weapons, even if shot down, can release their deadly payload in the air without the microbes being neutralised, thereby spreading throughout the area over the enemy.

It is not hard to imagine the military advantage on the battlefield of a drone attack over enemy lines to prepare for a conventional assault. Such a biological weapon attack would be surprising and operate under the enemy's radar until outbreaks of disease occur. The biological weapon used does not even need to be deadly; merely incapacitating a certain number of enemy soldiers through illness would suffice.

The rapid development of artificial intelligence will also likely have significant implications in the field of drones. So far, remotely piloted drones have been used. But drones can also be equipped with AI that makes them more autonomous in their missions.

Currently, military researchers in several countries, including the USA and China, are developing so-called drone swarms.[233] This refers to a large number of drones that can 'communicate' and interact through AI while possessing sensors that respond to their physical environment with various measures, such as selecting attack targets. With such drone swarms, much more can be achieved than with today's drones, which must be controlled individually. Although the development in this area is currently driven largely by military purposes, drone swarms may also prove useful in civilian applications.

Bioterrorists may also enhance their firepower by using drones carrying biological weapons, thereby increasing their dissemination over a larger area. In China, criminal gangs have reportedly already spread the swine influenza virus using drones to reduce the swine population in certain regions.[232, 233] The aim has been to scare farmers into selling their animals at reduced prices. It is likely only a matter of time before we witness bioterror attacks carried out with drones. It takes little imagination to envision the consequences of a drone attack with a deadly microbe dispersed as an aerosol over a large crowd, such as at a sports stadium with thousands of spectators gathered.

An Inevitable Evolution Towards More Dangerous Biological Weapons

The arsenal of biological weapons that existed 40–50 years ago would have caused extensive illness and death if deployed in military conflicts. Given the rapid developments we have seen more recently across various technology and research fields, it would be outright ostrich politics to deny that the dangers from biological weapons have increased significantly. It is imperative that this recognition leads to concrete consequences. We need well-thought out, updated preparedness plans for both preventing and combating attacks involving biological weapons.[234]

Chapter 17

The Road to Armageddon? – Thoughts on the Future

Among the feared group of CBRN weapons – chemical, biological nuclear, radioactive, and weapons – it is the threat of nuclear weapons that has dominated public discourse since the end of World War II. Chemical weapons, introduced during World War I but never used in World War II, have scarcely influenced public debate. This is even more true for biological weapons, despite the fact that several countries have had and still have such weapons. All CBRN weapons are referred to as weapons of mass destruction. Nevertheless, there is a widespread belief that it is only nuclear weapons that can trigger a global catastrophe and, in the worst-case scenario, lead to the extinction of humanity.

In reality, one must acknowledge that the use of biological weapons could have consequences as catastrophic as those caused by nuclear weapons. This is evident when comparing the damage inflicted by a biological attack with the consequences of a nuclear attack on a city the size of Oslo. If a 100-kiloton TNT bomb, which is five to six times stronger than the bombs dropped on Hiroshima and Nagasaki, were detonated over Oslo, new calculations suggest that this would result in over 100,000 deaths and nearly 190,000 injured.[235] An attack on the city with anthrax spores spread from an aeroplane could kill 95,000 and injure 125,000 according to 50-year-old WHO estimates.[236] With newer and more dangerous biological weapons likely to be developed, these numbers could be significantly higher. An anthrax attack would not lead to further spread via person-to-person transmission, but this could happen with other types of biological weapons, increasing the impact over a much larger area than the initial strike zone.

If the use of nuclear weapons in war could lead to the annihilation of humanity – an Armageddon as described in the Book of Revelation in the

Bible – it would require almost global use of such weapons. In contrast, a single attack with a highly lethal microbe that spreads through infection could trigger a pandemic that could wipe out humanity. There are therefore several paths to Armageddon, and the use of biological weapons is one of them. As we remember, the plague on its pale horse is one of the four horsemen of the apocalypse.

The Covid-19 pandemic illustrates uncomfortably clearly what consequences even a pandemic microbe with low lethality can have. The influenza virus that caused the Spanish flu 100 years ago resulted in the deaths of between 50 and 100 million people.[13] If microbes with significantly higher virulence, such as a bird flu virus – genetically manipulated as a biological weapon with very high transmissibility – were to spread, it would trigger a catastrophe. While bird flu viruses have a lethality of at least 60%, modern synthetic biology can likely produce viruses with much higher lethality rates. The prospects are frightening.

Ethnic or genetically selective biological weapons, as we have already touched upon, are no longer science fiction but may become a reality in the near future. This raises horrifying perspectives: we could be facing genocide.

The most extensive catastrophes that may occur after the use of biological weapons will likely result from state use of such weapons, since primarily states have the necessary resources to develop the most sophisticated biological weapons and effective dissemination methods using aircraft, missiles, artillery, and larger drones. However, as mentioned earlier, the possibilities for effective bioterrorism have also increased significantly, particularly with drone technology and artificial intelligence. This means that bioterror attacks could also have very serious consequences – and, in the worst case, trigger a pandemic if the microbe used possesses the necessary transmissibility characteristics.[237]

Biological weapons against plants and animals have, as we recall, also formed part of many states' biological weapons programmes, including in the Soviet Union and the USA. Targeted use of such weapons in a military conflict could lead to significant problems for the attacked party, severely disrupting the food supply. The 'anthrax cakes' that Britain developed during World War II were intended to decimate the cattle stock in Germany, thus creating famine.

The use of biological weapons against animals and plants can thus have very serious ecological consequences, with loss of biodiversity and, in the worst case, ecocide.[121]

It is both surprising and alarming that neither the general public nor politicians seem concerned about biological weapons and the very serious consequences their use could have – potentially global catastrophes. The fact that the threat from nuclear weapons has dominated public discourse has led to the establishment of several organisations aimed at combating the use of such weapons. When will we see similar organisations against biological weapons?

The striking silence regarding biological weapons may be due to a naive belief that the traditional aversion to them – an aversion enshrined in the BWC – prevents states or individuals from pursuing such weapons. We have seen numerous examples of distinguished scientists, motivated by various factors – patriotism, ideological fanaticism, and sheer opportunism – using their scientific expertise to develop deadly biological weapons. This includes doctors who forget the ancient Hippocratic oath that prohibits them from using medicine to harm others.[238, 239]

As we have seen, the explosive development in biotechnology, synthetic biology, and genetics is closely intertwined with the increasing risk of new and far more dangerous biological weapons in the years to come. This is perhaps one of the biggest challenges we face in the fight against biological weapons. The rapid advancement in biotechnology is driven by several factors that are difficult to control individually.

First and foremost, there are vast economic interests in modern biotechnological research, driven both by numerous commercial actors and academic institutions, where many scientists are focused on patents and profit opportunities.[240] Often, there are alliances between commercial companies and academic scientists who have established their own firms to conduct their research. With the enormous economic interests involved, it is evident that it is challenging to regulate this sector. That ethics often takes a backseat in this context is hardly surprising. Modern biotechnology and synthetic biology offer golden promises of solutions to numerous societal problems – both within and outside of medicine.

At certain religious festivals in Hinduism, images of the god Vishnu are taken out on an enormous cart – a 'juggernaut' – pulled by believers. It has been said that they not infrequently throw themselves into ecstasy under the wheels and are crushed. Modern biotechnology has been compared to a juggernaut rolling forth and crushing ethical objections and hesitations beneath its powerful wheels.[241] Influential and articulate advocates of this enterprise often effectively oppose oversight and regulation of projects, without extensive and well-informed debates about potential risks.

In addition, there is often an intimate collaboration between the aforementioned civilian actors and military research institutions regarding bio-defence. In his farewell address in 1961, President Dwight D Eisenhower warned against what he called the military-industrial complex – a powerful alliance between military leaders and industry in the development and production of weapons, wherein activities were largely driven by a scientific-technological elite with disproportionately large political influence.[242]

Following Amerithrax – the anthrax attack in 2001 – investments in bio-defence in the USA were increased significantly, and there was a large investment in civilian research environments. This led to parts of biotechnological research and its entrepreneurs, both within and outside academia, becoming a central part of the military-industrial complex. Today, we can indeed speak of a military-bio-industrial complex driven by military, political, and commercial motives. This complex is now more powerful than ever.[243]

As we have seen from many examples, the organisational blending of military and civilian research environments can also provide a breeding ground for more nefarious activities, such as the development of biological weapons under the guise of bio-defence. Biopreparat in the Soviet Union is a historical example that illustrates how the active development of biological weapons can effectively be camouflaged as civilian research. We have seen the same in South Africa's biological weapons programme, Project Coast. Regardless, the issues related to the dual-use dilemma and gain-of-function research will be significant in both military and civilian research related to bio-defence. Insight and reassuring oversight of this activity are therefore essential to establish both nationally and internationally.[244,245] An important part of this work must be to strengthen the international Biological Weapons Convention (BWC).

The horrifying consequences of the atomic bombings of Hiroshima and Nagasaki in 1945 led many scientists who had worked on the Manhattan Project that developed these bombs to have serious reservations. They now warned against further investment in nuclear weapons. This included the legendary leader of the project, J Robert Oppenheimer, who stated that 'physicists have known sin'.[246] The so-called 'Oppenheimer moment' refers to the moment when this central scientist recognised the potentially catastrophic consequences of the research to which he had passionately devoted himself.

Similar 'confessions' have never come from scientists who have been central to the development of biological weapons. In the heated, ongoing debate about the control of potentially dangerous gain-of-function research, one does not get the impression that scientists engaged in this activity are concerned about the risks. As a result, many of them fight tooth and nail against restrictions on their research, which is now part of the military-bio-industrial complex.[234, 241]

Should not leading scientists in the ever-expanding field of biotechnological research speak out against the threats posed by this research when it goes astray? Should they not advocate for what the author Richard Preston has called 'the taming of the biological beast'?[247] Such an echo from the nuclear physicists' warnings against nuclear weapons a couple of generations ago would hopefully awaken responsible politicians and lead to more powerful measures against the development of biological weapons and risky research. The risk is that so-called 'life sciences' could end up as 'death sciences'.

Oppenheimer was so overwhelmed by the sight of the first atomic bomb explosion in the Nevada desert that he utilised his Sanskrit knowledge to describe the immense power with a quote from the ancient Hindu epic, the Bhagavad Gita, where the god Vishnu says: 'I am death and the destroyer of worlds.'[248] The remarkable possibilities that biotechnological research holds today, where it is possible not only to alter existing organisms but also to create entirely new ones, can also evoke mythological associations. I referred previously to the Frankenstein motif in Mary Shelley's classic from 1818. The full title of the book was *Frankenstein; Or The Modern Prometheus*, where the author compared the bold and unrestrained researcher Frankenstein with the figure of Prometheus in Greek mythology.

Prometheus was the rebel who stole the sacred fire from the sky god Zeus and used it to give the spark of life to the humans he had created from clay. Zeus initially punished Prometheus by sending him the beautiful and beguiling woman Pandora with a box that she was to bestow upon him. When the box was opened, numerous misfortunes and disasters flew out into the world. Prometheus was ultimately punished by being chained to a mountain in the Caucasus, where an eagle constantly pecked at his liver.

The 'father' of the atomic bomb, Oppenheimer, has been compared to Prometheus. Furthermore, modern biotechnological research with Promethean ambitions, unrestrained and unopposed, could open a Pandora's box of disasters – new frightening biological weapons and pandemics.

Yet the Greek myth also provides grounds for optimism: At the bottom of Pandora's box – after the disasters had flown out – hope remained. There is still hope that humanity can tame 'the biological beast' through effective national and international measures. But this requires that one first acknowledges the problems. In this way, we can avoid a pandemic Armageddon.

Notes

1. Jansen HJ et al. 'Biological warfare, bioterrorism, and biocrime', Clin Microbiol Infect 2014; 20: 488–496.
2. Borio LL, Hynes NA, Henderson DA. 'Bioterrorism: An Overview', pp.3951–3963 in Mandell GL, Bennett JE, Dolin R (eds). Principles and Practice of Infectious Diseases. 7th ed. Churchill Livingstone Elsevier: Philadelphia, 2010.
3. Christian MD. 'Biowarfare and Bioterrorism', Crit Care Clin 2013; 29: 717–756.
4. Riedel S. 'Biological warfare and bioterrorism: a historical review', Proc (Bayl Univ Med Cent). 2004; 17: 400–406.
5. Lockwood JA. 'Insects as Weapons of War, Terror, and Torture', Ann Rev Entomol 2012; 57: 205–227.
6. Mayor A. Greek Fire, Poison Arrows, and Scorpion Bombs. Unconventional Warfare in the Ancient World. Princeton University Press: Princeton and Oxford, 2022.
7. Carus WS. 'The History of Biological Weapons Use: What We Know and What We Don't', Health Security 2015: 13: 219–255.
8. Trevisano SI. 'The 'Hittite plague', an epidemic of tularemia and the first record of biological warfare', Medical Hypotheses 2007; 69: 1371–1374.
9. Janowitz N. 'Inventing the Scapegoat: Theories of Sacrifice and Ritual', Journal of Ritual Studies. 2011; 25: 15–24.
10. Rolle R. The World of the Scythians. Batford: London, 1989. Oslo.
11. Turner MD. 'Mad Honey and the Poisoner King: A Case of Mass Grayanotoxin Poisoning in the Roman Military', Cureus 15(4): e38289. doi: 10.7759/cureus.38289.
12. Wheelis M. 'Biological Warfare at the 1346 Siege of Caffa', Emerging Infectious Diseases 2002; 8: 971–975.
13. Frøland SS. Duel Without End - Mankind's Battle with Microbes. Reaktion Books Ltd: London, 2022
14. Frandsen K-E. The Last Plague in the Baltic Region 1709–1713, Museum Tusculaneum Press: Copenhagen, 2010.
15. Thalassinou E et al. 'Biological Warfare Plan in the seventeenth century – the Siege of Candia, 1648–1669', Emerging Infectious Diseases 2015; 21: 2148–2153.
16. McNeill WH. Plagues and Peoples. Anchor Press/Doubleday: New York, 1976.
17. Carus WS. A Short History of Biological Warfare: From Pre-History to the 21st Century. National Defense University Press: Washington, DC, 2017.
18. Levi-Strauss C. Tropical Elegy. Gyldendal Norsk Forlag: Oslo, 1973.
19. Thorpe N, Willis O, Willis O, Smith C. '"Devil devil" – The sickness that changed Australia', ABC News, 21 June 2023.
20. Sartin JS. 'Infectious diseases during the civil war: The triumph of the "Third army"', Clin Infect Dis 1993; 16: 580–584.

21. Sitter N. The History of Terrorism – Assassination and Anti-Terror Efforts from Bakunin to IS. Dreyer Publishers: Oslo, 2017.
22. Costa H, Baños J-E. 'Bioterrorism in the literature of the nineteenth century: The case of Wells and The Stolen Bacillus', Cogent Arts & Humanities. (2016), 3: 1224538, https://doi.org/10.1080/23311983.2016.1224538.
23. London J. The Unparalleled Invasion. McClure's Magazine, July, 1910.
24. Geissler E. 'Biological warfare activities in Germany, 1923–1945', pp.91–126 in Geissler E, van Courtland Moon JE (eds). Biological and Toxin Weapons: Research, Development and Use from the Middle Ages to 1945. Oxford University Press: Oxford, 1999.
25. Lepick O. 'French activities related to biological warfare, 1919–1945,' pp.70–90 in Geissler E, van Courtland Moon JE (eds). Biological and Toxin Weapons: Research, Development and Use from the Middle Ages to 1945. Oxford University Press: Oxford, 1999.
26. Otto von Rosen. Wikipedia, 2023.
27. Harris S. 'The Japanese biological warfare programme, an overview', pp.127–152 in Geissler E, van Courtland Moon JE (eds). Biological and Toxin Weapons: Research, Development and Use from the Middle Ages to 1945. Oxford University Press: Oxford, 1999.
28. Mangold T, Goldberg J. Plague Wars – A True Story of Biological Warfare. Pan Books: London, 1999.
29. Cox SJ. H-057-2: 'I-400 and Operation Cherry Blossoms at Night: Japanese Plan for Biological Warfare – September 1945', Naval History and Heritage Command, January 2021.
30. King W, Guillemin J. 'The price of alliance: Anglo-American intelligence cooperation and Imperial Japan's criminal biological warfare programme, 1944–1947', Intelligence and National Security 2019; 34: 263–277.
31. Leitenberg M, Zilinskas RA with Kuhn JH. The Soviet Biological Weapons Programme – A History. Harvard University Press: Boston, 2012.
32. Bojtzov V, Geissler E. 'Military biology in the USSR, 1920–1945', pp.153–167 in Geissler E, van Courtland Moon JE (eds). Biological and Toxin Weapons: Research, Development and Use from the Middle Ages to 1945. Oxford University Press: Oxford, 1999.
33. Alibek K, with Handelman S. Biohazard – The Chilling True Story of the Largest Covert Biological Weapons Programme in the World – Told from the Inside by the Man Who Ran It. Random House: New York, 1999.
34. Guillemin J. Biological Weapons – From the Invention of State-Sponsored Programmes to Contemporary Bioterrorism. Columbia University Press: New York, 2005.
35. Kurt Blome. Wikipedia, 2023.
36. Reinhardt K. 'The Entomological Institute of the Waffen-SS: Evidence for Offensive Biological Warfare Research in the Third Reich', Endeavour 2013; 37: 220–227.
37. Kulik RM. 'Project Paper Clip – United States Government Programme', Britannica, 2023.
38. Carter GB, Pearson GS. 'British biological warfare and biological defense, 1925–1945', pp.168–189 in Geissler E, van Courtland Moon JE (eds). Biological and Toxin Weapons: Research, Development and Use from the Middle Ages to 1945. Oxford University Press: Oxford, 1999.

39. Rogers K. 'Sir Frederick Grant Banting – Canadian physician', Britannica, 2023.
40. Barnaby W. The Plague Makers – The Secret World of Biological Warfare. Vision Paperbacks: London, 1997.
41. Balmer B. Britain and Biological Warfare – Expert Advice and Science Policy, 1930–1965. Palgrave: Houndmills, Basingstoke, Hampshire, 2001.
42. Mahlen SD. 'Serratia infections: from military experiment to current practice', Clin Microbiol Rev 2011; 24: 755–791.
43. Leonard A. Cole. Clouds of Secrecy – The Army's Germ Warfare Tests over Populated Areas. Roman & Littlefield: Boston, 1988.
44. Balmer B, Spelling A, McLeish C. 'Preventing 'A Virological Hiroshima': Cold War Press Coverage of Biological Weapons Disarmament', Journal of War & Culture Studies 2016; 1: 74–90.
45. Seymour M Hersh. Chemical and Biological Warfare – America's Hidden Arsenal. Bobbs-Merrill: Indianapolis, 1968.
46. Fenton B. 'Trawler steamed into germ warfare site and no one said a word', The Telegraph, 20 September 2005.
47. Zilinskas RA. 'The Soviet Biological Weapons Programme and Its Legacy in Russia', Center for the Study of Weapons of Mass Destruction. Occasional Paper, No. 11, National Defense University Press: Washington, D.C., 2016.
48. Ings S. Stalin and the Scientists. A History of Triumph and Tragedy 1905–1953. Faber and Faber: London, 2016.
49. Roll-Hansen N. The Lysenko Effect: The Politics of Science. Humanity Books, Amherst, N.Y., 2005.
50. Tucker JB. 'The body's own bioweapons', Bulletin of the Atomic Scientists 2008; 64: 16–22 and 56–57, DOI: 10.1080/00963402.2008.11461128.
51. Gilsdorf JR. 'Commentary on the 1971 Smallpox Epidemic in Aralsk, Kazakhstan, and the Soviet Biological Warfare Programme (No. 2)', Crit Rev Microbiol 2003; 29: 163–167.
52. Miller J, Engelberg S, Broad W. Germs – The Ultimate Weapon. Simon and Schuster: New York, 2001.
53. Martinez I. 'The History of the Use of Bacteriological and Chemical Agents during Zimbabwe's Liberation War of 1965–80 by Rhodesian Forces', Third World Quarterly 2002; 23: 1159–1179.
54. Nass M. 'Epozootic in Zimbabwe, 1978–80: due to deliberate spread?', PSR Quarterly 1992; 2: 198–209.
55. Wilson JM et al. 'Reanalysis of the anthrax epidemic in Rhodesia, 1978–1984', PeerJ 2016 Nov 10; 4: e2686. doi: 10.7717/peerj.2686. PMID: 27867766; PMCID: PMC5111893.
56. Purkitt HE, Burgess S. 'South Africa's Chemical and Biological Warfare Programme: A Historical and International Perspective', Journal of Southern African Studies 2002; 28: 229–253.
57. Gould C, Folb P. 'The role of professionals in the South African chemical and biological warfare programme', Minerva 2002; 40: 77–91.
58. Singh JA. 'Project Coast: Eugenics in Apartheid South Africa', Endeavour 2008; 32: 5–9.
59. Jackson MA. 'Conspiracy to Commit Genocide – Anti-Fertility Research in Apartheid's Chemical and Biological Weapons Programme', Journal of International Criminal Justice 2015; 13: 933–950.

60. Marbot O. South Africa: '"Dr. Death" discovered to be still practicing medicine', The Africa Report, 5 February 2021.
61. Morris B, Kedar BZ. "Cast thy bread': Israeli biological warfare during the 1948 war', Middle Eastern Studies 2023; 59: 752–776.
62. Cohen A. 'Israel and Chemical/Biological Weapons: History, Deterrence, and Arms Control', The Nonproliferation Review 2001; 8: 27–53.
63. Tamkin E. 'Colin Powell did not want the Iraq War to define his legacy – but it will', The New Statesman, 18 October 2021.
64. Andrew Grice. 'Whitewash: The David Kelly Saga that Ended in Tragedy and Tarnished New Labour Forever', Independent, 19 March 1923.
65. Lister S. 'Doctors claim cover-up over the death of weapons expert Dr. David Kelly', Belfast Telegraph, 17 July 2013.
66. Burghardt T. 'Dick Cheney's "Executive Assassination Ring". Was British Weapons Expert Dr. David Kelly a Target?', Global Research, 17 July 2009.
67. Ainscough MJ. 'Next Generation Bioweapons: The Technology of Genetic Engineering Applied to Biowarfare and Bioterrorism', Counterproliferation Paper No. 14, 2002.
68. Petersen R. 'Fear and loathing in Moscow – The Russian biological weapons programme in 2002', Bulletin of the Atomic Scientists, 5 October 2022.
69. Tucker JB. 'Biological weapons in the former Soviet Union. An interview with Dr. Kenneth Alibek', The Nonproliferation Review/Spring/Summer 1999.
70. Shoham D, Wolfson Z. 'The Russian Biological Weapons Program: Vanished or Disappeared?', Crit Rev Microbiol 2004; 30: 241–261.
71. Preston R. The Cobra Event. Random House: New York, 1998.
72. Wheelis M, Dando M. 'Back to bioweapons? The United States may have rejected the bioweapons protocol because it is committed to continuing and expanding its secret programmes', Bulletin of the Atomic Scientists 2003; 59: 1–6.
73. Reeves RG et al. 'Agricultural research, or a new bioweapon system?', Science 2018; 362: 35–37.
74. Harris ED. 'North Korea and Biological Weapons: Assessing the Evidence', 38 North Special Report, Stimson, 6 November 2020.
75. Baumgaertner E, Broad WJ. 'North Korea's Less-Known Military Threat: Biological Weapons', The New York Times, 15 January 2019.
76. Kania EB, Vorndick W. 'Weaponising Biotech: How China's Military is Preparing for a "New Domain of Warfare"', Defense One, 14 August 2019.
77. Aryan J. 'A look at China's biowarfare ambitions', ORF, 2 June 2021.
78. Bansak KC. 'Managing networks of risk: A tailored approach to Iran's biological warfare threat potential', Bulletin of the Atomic Scientists 2011; 67: 66–76.
79. Carus WS. 'Iranian Nuclear, Biological and Chemical Weapons: Implications and Responses', Middle East Review of International Affairs 1998; 2: 12–22.
80. Carus WS. 'Defining "Weapons of Mass Destruction"', Center for the Study of Weapons of Mass Destruction. Occasional Paper, No. 8. National Defense University Press: Washington, D.C; Washington, 2012.
81. Spiers EM. A History of Chemical and Biological Weapons. Reaktion Books: London, 2010.
82. Reiners S, Thränert O. 'The Biological Weapons Ban and Scientific Progress', CSS Analyses in Security Policy, No. 321. Center for Security Studies, ETH Zurich, April 2023.

83. Kirby J. 'Can a 50-year-old treaty still keep the world safe from the changing threat of bioweapons?', Vox, 28 April 2023.
84. Naik S, Ramanathan A. 'The New Bioweapons Peril: A case to revisit the Biological Weapons Convention', Indian Public Policy Review 2022; 3: 59–76.
85. Saunders E. 'Can the Biological Weapons Convention address new biothreats?', Chatham House, 25 November 2021.
86. Carus WS. 'Bioterrorism and Biocrimes – The Illicit Use of Biological Agents Since 1900', Center for Counterproliferation Research, National Defense University: Washington, D.C., 1998 (February 2001 Revision).
87. Carus WS. 'R.I.S.E', pp.55–70 in Tucker JB (ed) Toxic Terror – Assessing Terrorist Use of Chemical and Biological Weapons. MIT Press: Cambridge, Mass., 2000.
88. Lindholm C. 'Culture, Charisma, and Consciousness: The Case of the Rajneeshee', Ethos 2002; 30: 357–375.
89. Carus WS. 'The Rajneeshees', pp.115–137 in Kaplan DE, Tucker JB (eds) Toxic Terror – Assessing Terrorist Use of Chemical and Biological Weapons. MIT Press: Cambridge, Mass., 2000.
90. Tucker JB, Pate J. 'The Minnesota Patriots Council', pp.159–183 in Tucker JB (ed) Toxic Terror – Assessing Terrorist Use of Chemical and Biological Weapons. MIT Press: Cambridge, Mass., 2000.
91. Kaplan DE. 'Aum Shinrikyo', pp.207–226 in Tucker JB (ed) Toxic Terror – Assessing Terrorist Use of Chemical and Biological Weapons. MIT Press: Cambridge, Mass., 2000.
92. Bleek PC. 'Revisiting Aum Shinrikyo: New Insights into the Most Extensive Non-State Biological Weapons Programme to Date', NTI, 10 December 2011.
93. Danzig R et al. Insights Into How Bioterrorists Develop Biological and Chemical Weapons. 2nd ed. Center for a New American Security, 2012.
94. Tanimoto T et al. 'Involvement of doctors in Aum Shinrikyo', The Lancet 2018; 392: 1116.
95. Murakami H. Underground: The Tokyo Gas Attack and the Japanese Psyche. Vintage: London, 2003.
96. Lifton RJ. Destroying the World To Save It: Aum Shinrikyo, Apocalyptic Violence, and the New Global Terrorism. Picador: London, 2000.
97. Gunaratna R. 'Aum Shinrikyo's Rise, Fall and Revival', Counter Terrorist Trends and Analyses 2018; 10: 1–6.
98. Baffelli E. 'Did Aum Shinrikyo Really End?', pp.49–66 in Stausberg M, Wright SA, Cusack CM (eds). The Demise of Religion: How Religions End, Die, or Dissipate. Bloomsbury Academic: London, 2020.
99. Cole LA. 'The US Anthrax Letters: A Confirmed Case of BW Agent Use', pp.23–48 in Clunan AL, Lavoy PR, Martin SB (eds) Terrorism, War, or Disease? Stanford Security Studies: Stanford, Calif., 2008.
100. Willman D. The Mirage Man: Bruce Ivins, the Anthrax Attacks, and America's Rush to War. Bantam: New York, 2011.
101. Gjerding ML, Helsingen T, Utheim EB. 'Breivik considered biological weapons; now terror preparedness is increasing: Preparing for bio-terror in Norway', VG 12 November 2012.
102. Hellmich C, Redig AJ. 'The Question is When: The Ideology of Al-Qaeda and the Reality of Bioterrorism', Studies in Conflict & Terrorism 2007; 30: 379–396.
103. 'Conversation with Terror', Time Magazine, 11 January 1999.

104. Kellman B. Bioviolence – Preventing Biological Terror and Crime. Cambridge University Press: Cambridge, 2007.
105. Tomlinson C. 'Iranian Bio-Bomber Convicted in German Court', The European Conservative, June 22, 2023.
106. Segel L. '"The yellow fever plot" – Germ warfare during the Civil War', Canadian Journal of Diagnosis, September 2002.
107. Campbell D. 'Close but no cigar: how America failed to kill Fidel Castro', The Guardian, 26 November 2016.
108. Hayes, Amy. 'How the CIA Attempted to Assassinate Fidel Castro', TheCollector.com, 29 March 2023, www.thecollector.com/how-the-cia-attempted-to-assassinate-fidel-castro.
109. Catanzaro M. 'Was famed poet Pablo Neruda poisoned? Scientists warn case not closed', Nature 2023; 615: 18–19.
110. Whiteside A. 'HIV/AIDS – A Very Short Introduction', Oxford University Press: Oxford, 2008.
111. Colgrass N. 'Son, 24, forgives dad who injected him with HIV: "That's all I can do"', USA Today, 3 May 2016.
112. Spencer RC, Lightfoot NF. 'Preparedness and Response to Bioterrorism', Journal of Infection 2001; 43: 104–110.
113. 'Chemical and Biological Warfare Overview 2015', United States Air Force, Center for Unconventional Weapons Studies, 2 January 2015.
114. Adalja AA. 'Identifying and managing casualties of biological terrorism', UpToDate, version January 2024.
115. Green MS et al. 'Confronting the threat of bioterrorism: realities, challenges, and defensive strategies', The Lancet 2019; 19: e2–13.
116. Frinking E et al. 'The increasing threat of biological weapons – Handle with sufficient and proportionate care', The Hague Centre for Strategic Studies: The Hague, 2016.
117. 'Singapore: On alert for airborne threats – Sensor to detect biological agents in air on trial at land checkpoint', Aseanplus News, 6 February 2023.
118. Mullin E. 'How to Detect a Man-Made Biothreat', WIRED, 1 November 2022.
119. Hilleman MR. 'Overview: cause and prevention in biowarfare and bioterrorism', Vaccine 2002; 20: 3055–3067.
120. O'Toole T, Mair M, Inglesby TV. 'Shining light on "Dark Winter"', Clin Infect Dis 2002; 34: 972–983.
121. Dudley JP, Woodford MH. 'Bioweapons, Biodiversity, and Ecocide: Potential Effects of Biological Weapons on Biological Diversity', Bio Science 2002; 52: 583–592.
122. Olson D. 'Agroterrorism – Threats to America's Economy and Food Supply', FBI Law Enforcement Bulletin 2012; 81: 1–10.
123. Radosavljevic V. 'Environmental Health and Bioterrorism', 2019; 2: 450–57.
124. Crutchley TM et al. 'Agroterrorism: Where Are We in the Ongoing War on Terrorism?' Journal of Food Protection 2007; 70: 791–804.
125. Keremidis H et al. 'Historical Perspective on Agroterrorism: Lessons Learned from 1945 to 2012', Biosecurity and Bioterrorism: Biodefense Strategy, Practice, and Science 2013; 11, Suppl 1: S17–S24.
126. Yeh J-Y et al. 'Livestock Agroterrorism: The Deliberate Introduction of a Highly Infectious Animal Pathogen', Foodborne Pathogens and Disease 2012; 9: 869–877.

127. Madden LV, Wheelis M. 'The Threat of Plant Pathogens as Weapons Against U.S. Crops', Annu Rev Phytopathol 2003; 41: 155–176.
128. Tucker JB. 'Introduction', pp.1–16 in Tucker JB (ed). Innovation, Dual Use, and Security – Managing the Risks of Emerging Biological and Chemical Technologies. The MIT Press: Cambridge, Mass., 2012.
129. Lentzos F. 'Dual Use in Biology and Biomedicine', Nuffield Council on Bioethics. November 2015 (last updated on 28 July 2022).
130. Young A. Pandora's Gamble. Lab Leaks, Pandemics, and a World at Risk. Center Street: Nashville/New York, 2023.
131. Kuiken T. 'Oversight of Gain of Function Research with Pathogens: Issues for Congress', Congressional Research Service, 26 May 2022.
132. Ebright R. 'Written testimony. Submitted for the Record to the US Senate Committee on Homeland Security, Subcommittee on Emerging Threats and Spending Oversight', 3 August 2022.
133. Klotz L. 'The grave risk of lab-created potentially pandemic pathogens', Bulletin of the Atomic Scientists, 9 September 2021.
134. Musunuri S et al. 'Rapid Proliferation of Pandemic Research: Implications for Dual-Use Risks', Perspective 2021; 12: e0 1864–1821.
135. Wain-Hobson S. 'Gain-of-function research can't deliver pandemic predictions. Are there alternatives?' Bulletin of the Atomic Scientists, 27 June 2022.
136. Biosafety Now!, http://biosafetynow.org.
137. Goodrum F et al. 'Virology under the Microscope – a Call for Rational Discourse', Journal of Virology 2023; 97 Issue 2 e00089-23.
138. Field M. 'Gain-of-function pathogen research is controversial and widespread. Can it be regulated?' Bulletin of the Atomic Scientists, 25 September 2023.
139. Reardon S. 'Stricter US guidelines for 'gain-of-function' research are on the way – maybe', Nature, 31 January 2023.
140. Bayot BL, King KC. Biohazard Levels. National Center for Biotechnology Information, US National Library of Medicine, Bethesda, Md. 2019.
141. Klotz L. 'Human error in high-biocontainment labs: a likely pandemic threat', Bulletin of the Atomic Scientists, 25 February 2019.
142. Blacksell SD et al. 'Laboratory-acquired infections and pathogen escapes worldwide between 2000 and 2001: a scoping review', The Lancet Microbe, 12 December 2023. DOI: https://doi.org/10.1016/S2666-5247(23)00319-1.
143. Rodgers J et al. 'How to make sure the labs researching the most dangerous pathogens are safe and secure', Bulletin of the Atomic Scientists, 2 July 2021.
144. Mallapaty S. 'Labs that handle dangerous pathogens surge in wake of Covid', Nature 2022; 610: 428–429.
145. Keuch GT et al. 'Pandemic origin and a "One Health" approach to preparedness and prevention: Solutions based on SARS-CoV-2 and other RNA viruses', Proc Natl Acad Sci U S A; 119(42): e2202871119, 2022 10 18.
146. Chan A, Ridley M. Viral – The Search for the Origin of Covid-19. 4th Estate: London, 2021.
147. Wade N. 'The origin of COVID: Did people or nature open Pandora's box at Wuhan?', Bulletin of the Atomic Scientists, 5 May 2021.
148. Paul R. Deception – The Great Covid Cover-Up. Regnery Publishing: Washington, D.C., 2023.

149. 'Potential Links Between the Wuhan Institute of Virology and the Origin of the COVID-19 Pandemic', Office of the Director of National Intelligence, 23 June 2023.
150. Eban K. '"This shouldn't Happen": Inside the Virus-Hunting Nonprofit at the Center of the Lab-Leak Controversy', Vanity Fair, 31 March 2022.
151. Relman DA. 'To stop the next pandemic, we need to unravel the origins of COVID-19', PNAS 2020; 117: 29246–29248.
152. Bloom JD et al. 'Investigate the origins of COVID-19', Science 2021; 372: 694.
153. Shir-Raz Y et al. 'Censorship and Suppression of Covid-19 Heterodoxy: Tactics and Counter-Tactics', Minerva, 1 November 2022.
154. Lima-Strong C. 'New report on Covid-19 origin puts social media in GOP's crosshairs', The Washington Post, 27 February 2023.
155. McCarthy C. 'Dr. Jeffrey Sachs Says COVID About "Massive Government Lying" Quits Democrats', Newsmax, 24 November 2023.
156. Fauci AS. 'Research on Highly Pathogenic H5N1 Influenza Virus. The Way Forward', mBio, 9 October 2012.
157. Mordowanec N. 'Fauci Had Secret Meeting With CIA About COVID, Republicans Claim', Newsweek, 27 September 2023.
158. Kennedy B, Tyson A. 'Americans' Trust in Scientists, Positive Views of Science Continue to Decline', Pew Research Center, 14 November 2023.
159. Kolodziejczyk B, Malone J. 'Do-it-yourself biology: an open innovation movement or a threat?', https://globalyoungacademy.net, May 2018.
160. Greene D, Pannu J, Berke A. 'The Danger of "Invisible" Biolabs Across the U.S.', Time, 31 August 2023.
161. Nava V. 'Pathogens labeled "HIV" and "Ebola" found inside secret, illegal Chinese-owned biolab in California', New York Post, 16 November 2023.
162. Leitenberg M. 'False Allegations of U.S. Biological Weapons Use During the Korean War', pp.120–123 in Clunan AL, Lavoy PR, Martin SB (eds) Terrorism, War, or Disease? Stanford Security Studies: Stanford, Calif., 2008.
163. Pick D. Brainwashed – A New History of Thought Control. Wellcome Collection: London, 2022.
164. Chen S. 'History of Three Mobilisations: A Reexamination of the Chinese Biological Warfare Allegations against the United States in the Korean War', The Journal of American-East Asian Relations 2009; 16: 213–247.
165. Buchanan T. 'The Courage of Galileo: Joseph Needham and the "Germ Warfare" Allegations in the Korean War', History 2001; 86: 503–522.
166. Regis E. The Biology of Doom – The History of America's Secret Germ Warfare Project. Henry Holt and Co: New York, 1999.
167. Leitenberg M. 'China's False Allegations of the Use of Biological Weapons by the United States during the Korean War', Wilson Center, March 2016.
168. Kuzmarov J. '"There's Something Rotten in Denmark": Frank Olson and the Macabre Fate of a CIA Whistleblower in the Early Cold War', Class, Race and Corporate Power 2020; Vol 8: iss 1, Article 3.
169. Kinzer S. Poisoner in Chief: Sidney Gottlieb and the CIA Search for Mind Control. Henry Holt and Co: New York, 2019.
170. Meeselson MS, Robinson JP. 'The Yellow Rain Affair: Lessons from a Discredited Allegation', pp.72–96 in Clunan AL, Lavoy PR, Martin SB (eds) Terrorism, War,

or Disease? Unraveling the Use of Biological Weapons Stanford Security Studies: Stanford, Calif., 2008.
171. Roberts J. 'The Mystery of Yellow Rain', Distillations Magazine, 13 April 2018.
172. Katz R. '"Yellow Rain" Biological Warfare Agent Use: Evidence and Remaining Questions', pp.97–119 in Clunan AL, Lavoy PR, Martin SB (eds) Terrorism, War, or Disease? Unraveling the Use of Biological Weapons Stanford Security Studies: Stanford, Calif., 2008.
173. Zilinskas R. 'Cuban Allegations of U.S. Biological Warfare: False Allegations and Their Impact on Attribution', pp.144–164 in Clunan AL, Lavoy PR, Martin SB (eds) Terrorism, War, or Disease? Unraveling the Use of Biological Weapons Stanford Security Studies: Stanford, Calif., 2008.
174. Lawrence K. 'New Evidence Implicates CIA in 1971 Attack on Cuba with African Swine Fever Virus', CovertAction Magazine, 5 October 2020.
175. Selvage D, Nehring C. 'Operation "Denver": KGB and Stasi Disinformation regarding AIDS', Wilson Center, 22 July 2019.
176. Kramer M. 'Lessons From Operation "Denver", the KGB's Massive AIDS Disinformation Campaign', The MIT Press Reader, 26 May 2020.
177. Boghardt T. 'Soviet Bloc Intelligence and Its AIDS Disinformation Campaign', Studies in Intelligence 2009; 53: 1–24.
178. Cohen J. 'AIDS Debate Declared Dead', Science, 25 April 2001.
179. Barrett R. 'The 1994 Plague in Western India: Human Ecology and the Risks of Misattribution', pp.49–71 in Clunan AL, Lavoy PR, Martin SB (eds) Terrorism, War, or Disease? Unraveling the Use of Biological Weapons Stanford Security Studies: Stanford, Calif., 2008.
180. Dutt AK, Akhtar R, McVeigh M. 'Surat plague of 1994 re-examined', Southeast Asian J Trop Med Public Health 2006; 37: 755–760.
181. WION Web Team. 'Found 30 biological labs in Ukraine, possibly for bioweapons claim Russian forces', WION, 8 March 2022.
182. Opindia Staff. 'Russian Defence Ministry says they found evidence of US-funded biolabs in Ukraine, deadly pathogens were "urgently destroyed"', Opindia, 7 March 2022.
183. Palmer E. 'U.S. Biological Weapons in Ukraine – Separating the Facts From the Fiction', Newsweek, 8 March 2022.
184. Field M. 'US official: Russian invasion of Ukraine risks release of dangerous pathogens', Bulletin of the Atomic Scientists, 25 February 2022.
185. Lanese N. 'Ukraine should destroy "high-threat" pathogens, WHO says', Live Science, 11 March 2022.
186. 'US biolabs need exposure and investigation: US scholar', Global Times, 21 March 2022.
187. Press TV. 'Kennedy Jr confirms existence of American biolabs in Ukraine', Press TV, 15 August 2022.
188. Field M. 'In Ukraine, US-military-linked labs could provide fodder for Russian disinformation', Bulletin of the Atomic Scientists, 9 March 2022.
189. Frøland SS. 'Bioweapons in Ukraine – propaganda or fact?' Foreign Policy, 1 April 2022.
190. McCarthy B. 'The facts behind the Russian right-wing narratives claiming Hunter Biden funded biolabs in Ukraine', Politifact, 1 April 2022.
191. Littlewood J, Lentzos F. 'Russia (again) peddles its debunked US-Ukrainian bioweapons claims at the United Nations', Bulletin of the Atomic Scientists, 15 November 2022.

192. Thornhill JP et al. 'Monkeypox Virus Infection in Humans across 16 Countries – April–June 2022', N Eng J Med 2022; 387: 679–691.
193. Stradner I. 'Kremlin Claims Monkeypox Could Be a Secret U.S. Bioweapon', Foreign Policy, 21 August 2022.
194. 'NTI Statement and FAQ Regarding NTI-Munich Security Conference 2021 Tabletop Exercise on Reducing High-Consequence Biological Threats', NTI, 24 May 2022.
195. Elsayed S, Bondy L, Hanage WP. 'Monkeypox Virus Infections in Humans', Clin Microbiol Rev 2022 Dec 21; 35(4): e0009222.
196. Frøland SS. 'Monkeypox and Conspiracy Theories', Foreign Policy, 4 September 2022.
197. Newey S. 'Monkeypox 'circulating for five years' before 2022's global health emergency', 3 November 2023.
198. Lee CT, Zhe D. 'Did the US ask Taiwan to develop a bioweapon targeting Chinese DNA?' Asia Fact Check Lab, Radio Free Asia, 1 August 2023.
199. Everington K. 'Taiwan building P4 lab for biosafety research, biowarfare defense', Taiwan News, 10 July 2023.
200. Field M, Krzyzaniak. 'Why do politicians push false bioweapon conspiracy theories on coronavirus?' People's World, 17 March 2020.
201. Helsel P. 'RFK Jr. pushes back on report he said Covid-19 was ethically targeted to spare Jews', NBC News, 15 July 2023.
202. Swaminathan S. 'COVID-19 was engineered as "bioweapon" to purposely infect people, claims Wuhan researcher', WION, 28 June 2023.
203. Tilley C. 'Chinese scientists "create" a mutant coronavirus strain that attacks the BRAIN and has a 100% kill rate in mice – as they admit there's a "risk that it spills over to humans"', Daily Mail, 18 January 2024.
204. Bentley M. 'What if China really did develop COVID as a bioweapon? Here are the issues involved', The Conversation, 14 June 2023.
205. Meld. St. 5 (2023–2024). 'A Resilient Public Health Preparedness – From Pandemic to War in Europe', Ministry of Health and Care Services, 24 November 2023.
206. McCauley PM, Payne RA. 'The Illogic of the Biological Weapons Taboo', Strategic Studies Quarterly 2010; 4: 6–35.
207. Alleslev L. 'Biological weapons: Technological progress and the spectre of bioterrorism in the post-Covid era', Preliminary draft report, SubCommittee on Technological Trends and Security. NATO Parliamentary Assembly, 22 March 2021.
208. Juling D. 'Future Bioterror and Biowarfare Threats for NATO's Armed Forces until 2030', Journal of Advanced Military Studies 2023; 14: 118–143.
209. Knapton S. 'Bill Gates: Terrorists could wipe out 30 million people by weaponising a disease such as smallpox', The Telegraph, 19 April 2017.
210. Turpeinen L. 'Engineering or God Delusion – What Exactly is Synthetic Biology?', Medium, 30 December 2019.
211. Ball P. 'Synthetic biology: starting from scratch', Nature 2004; 431: 624–626.
212. Matthews LJ et al. 'Plagues, Cyborgs, and Supersoldiers. The Human Domain of War', Rand Corporation, 2024.
213. National Academies of Sciences, Engineering, and Medicine. Biodefense in the Age of Synthetic Biology. The National Academies Press, Washington DC, 2018.
214. Jackson RJ et al. 'Expression of Mouse Interleukin-4 by a Recombinant Ectromelia Virus Suppresses Cytolytic Lymphocyte Responses and Overcomes Genetic Resistance to Mousepox', J Virol 2001; 75: 1205–1210.

215. Wimmer E. 'The test tube synthesis of a chemical called poliovirus: The simple synthesis of a virus has far-reaching societal implications', EMBO Rep. 2006 Jul; 7(Spec No): S3–S9.
216. Tumpey TM, et al. 'Characterisation of the reconstructed 1918 Spanish influenza pandemic virus', Science 2005; 310: 77–80.
217. Noyce RS, Lederman S, Evans DH. 'Construction of an infectious horsepox vaccine from chemically synthesized DNA fragments', PLoS One. 2018; 13(1): e0188453.
218. Gibson DG et al. 'Creation of a Bacterial Cell Controlled by a Chemically Synthesized Genome', Science 2010; 329: 52–66.
219. Heinlein RA. The Day after Tomorrow. Signet Books: New York, 1951.
220. Geraghty J. Hunting Four Horsemen – A Dangerous Clique Novel. Discus Books, 2020.
221. Appel JM. 'Is all fair in biological warfare? The controversy over engineered biological weapons', J Med Ethics 2009; 35: 429–432.
222. Lentzos F. 'How to protect the world from ultra-targeted biological weapons', Bulletin of the Atomic Scientists 2020; 76: 6, 302–308.
223. Geraghty J. 'The Coming Threat of a Genetically Engineered "Ethnic Bioweapon"', National Review, 10 April 2023.
224. Van Brugen I. 'Pentagon Making Race-Specific Bioweapons to Target Citizens, China Says', Newsweek, 11 May 2023.
225. Mcfarlane H. 'Why couldn't Q Save Bond? No Time To Die's Science Explained', SCREENRANT, 9 January 2023.
226. Burak A. 'The Promise and Challenges of Using Nanobots in Medicine', RELEVANT, 15 September 2023.
227. Hendrycks D, Mazeika M, Woodside T. 'An Overview of Catastrophic AI Risks', arXiv: 2306.12001 v2 (cs.CY).
228. Mouton CA, Lucas C, Guest E. 'The Operational Risks of AI in Large-Scale Biological Attacks. A Red-Team Approach', Rand Corporation, 2023.
229. Sandbrink J. 'ChatGPT could make bioterrorism horrifyingly easy', Vox, 7 August 2023.
230. Urbina F et al. 'Dual Use of Artificial Intelligence-Powered Drug Discovery', Nat Mach Intell 2022; 4: 189–191.
231. Kuiken T. 'Artificial Intelligence in the Biological Sciences: Uses, Safety, Security, and Oversight', Congressional Research Service, November 2023.
232. DeFranco J, Giordano J. 'Dark Side of Delivery: The Growing Threat of Bioweapons Dissemination by Drones', Defence iQ, 24 January 2020.
233. Nichols RK et al. 'Drone Delivery of CBNRECy-Dew Weapons of Mass Destruction and Disruption', Pressbooks: New York, 2022.
234. Yassif J. 'Preventing Catastrophic Bioterrorism: Guarding Against Exploitation of the Life Sciences and Biotechnology', CTC Sentinel, May 2022.
235. Henmo J. 'This is what a nuclear attack looks like', Aftenposten, 23 September 2023.
236. 'Report of WHO Group of Consultants. Health Aspects of Chemical and Biological Weapons', WHO, Geneva, 1970.
237. Webb A. 'The Next Pandemic Could Start with a Terrorist Attack', The Atlantic, 22 March 2023.
238. Zanders JP. 'Ethics and Reason in Chemical and Biological Weapons Research', Minerva 2002; 40: 3–13.

239. Guillemin J. 'Scientists and the history of biological weapons: A brief historical overview of the development of biological weapons in the twentieth century', EMBO Rep. 2006 Jul; 7(Spec No): S45–S49.
240. Krimsky S. Science in the Private Interest. Has the Lure of Profits Corrupted Biomedical Research? Rowman & Littlefield Publishers: Lanham, Maryland, 2003.
241. Stevens T, Newman S. Biotech Juggernaut – Hope, Hype, and Hidden Agendas of Entrepreneurial BioScience. Routledge: New York, 2019.
242. Leary WE. 'After 50 Years, Eisenhower's Warnings Against a Scientific Elite Still Cause Consternation', AAAS, 11 February 2011.
243. Hartung WD, Freeman B. 'The Military Industrial Complex is More Powerful Than Ever', The Nation, 9 May 2023.
244. Gronvall GK. 'Biotechnology Governance in a Time of New Risks and Opportunities', Council of Foreign Relations, 30 January 2023.
245. Rozpedowski J. 'Bio-Security In The Age Of Global Pandemics', Global Security Review, 17 July 2023.
246. Lifton RJ. 'Oppenheimer's tragedy – and ours', Bulletin of the Atomic Scientists, 17 July 2023.
247. Preston R. 'Taming The Biological Beast', The New York Times, 21 April 1998.
248. Blakemore E. 'Who is Oppenheimer? The controversial man behind the atomic bomb', National Geographic, 27 June 2023.

Index of Persons